THE BELLS
of
SAN FRANCISCO

THE BELLS
of
SAN FRANCISCO

THE SALVATION ARMY
WITH ITS SLEEVES ROLLED UP

By Judy Vaughn

The Salvation Army
Golden State Division
San Francisco, California

RDR Books
Berkeley, California

The Bells of San Francisco

RDR Books
2415 Woolsey Street
Berkeley, CA 94705
Phone: (510) 595-0595
Fax: (510) 228-0300
E-mail: read@rdrbooks.com
Website: www.rdrbooks.com

Copyright 2005 by The Salvation Army San Francisco
All Rights Reserved

ISBN: 1-57143-150-0

Library of Congress Control Number: 2005933785

Design and production: Richard Harris
Image Scanning: Proper Publishing
Printing Coordinator: Lynda Rea
Cover photo: Dorothea Lange, courtesy of the
Farm Security Administration collection "American Memory,"
Library of Congress

Distributed in Canada By Starbooks Distribution,
100 Armstrong Way, Georgetown, ON L7G 5S4

Distributed in the United Kingdom and Europe by
Roundhouse publishing Ltd., Millstone, Limers Lane,
Northam, North Devon EX39 2RG, United Kingdom

Printed in Canada by Transcontinental Printing

The Bells of San Francisco

INTRODUCTION: Telling the Army's Story   1
Anecdotes and Memories

ONE: Bringing the Booth Legacy to California   5
General William Booth and the Bay Area Press • Coming to California and Finding the Gold • A Selected Family Album—The Salvation Army in San Francisco

TWO: Disaster Relief   33
San Francisco in Ruins—"Greatest Catastrophe of the Century" • What Made Those Doughnuts Taste So Gosh Darn Good? • A Cup of Cold Water • Loma Prieta—Disaster in Our Own Back Yard

THREE: Drug Addiction   57
"Put Another Nickel on The Drum. Save a Drunken Bum." From the Very Beginning the Army Worked on Skid Row • "Falling Off the Wagon"—The Phrase Started with a Real Wagon and an Annual Parade • Please, Please Help My Son • Often Inherited, Always Developed by Indulgence, but Clearly a Disease • Moving Stuff—The Truck You Call Has More Than One Purpose

FOUR: Changing Times   74
101 Valencia Was the Soul of the Army's Administration, Worship, and Training • Overarching Principles

FIVE: The Act of Giving   86

SIX: Christmas and the Popular Image   89
Who's Making All That Noise? Volunteers Make Celebrity Bell Ringing Street Theater at its Finest • The Best of Times, the Worst of Times, the Pictures People Remember Most • At the Turn of the Century, Productions Like "Salvation Nell" Idealized The Salvation Army Lassie

SEVEN: South of Market   116
Changing the Skyline

EIGHT: Building Community   124
At the Heart of the Army

NINE: Who, What, When and Where   131
Among Those Who Have Served on the San Francisco Advisory Board • Councils and Special Committees • Northern California and Nevada Division • Golden State Division • Questions People Always Ask • International Mission Statement • For Further Information • Retrospective

Index

Acknowledgments

Lt. Colonel Chris Buchanan planted the seed for this book. Colonels Henry and Marie Koerner nurtured it. Lt. Colonels Richard and Bettie Love and then Majors Joe and Shawn Posillico made it possible.

With those beginnings and the friendship of Salvationists for almost 25 years, I approached the project with an arsenal of memories and a healthy respect for the history of this Army.

Many officers, soldiers and advisory board members have answered my questions and shared personal stories. Evie Dawe has been my mainstay, along with Pat Eberling and Evie Dexter who checked social service facts. And every time I see Major Judy Smith smile, I think I must be doing something right.

Objectively and with a sure hand, advisors from both inside and outside of the Army have been a strong support, reminding me that—in addition to being correct—history ought to be lively and readable.

For that I sincerely thank my family and friends including: Nancy Randall, Roz Leiser, Connell Persico, Gayle Leyton, Joe Reed Pierce and the Osher Lifelong Learning Institute at San Francisco State University; Bill and Laura Waste, Ernie and Gertrude Marx, Lily Chin, Sue Tom, Lyle Richardson, Lt. Colonels Charles and Elnora McIntyre, Lt. Colonel Check Hung Yee; Major Terry Camsey, Barry Frost and Ken Bricknell from southern California; Michael Svanevich from San Mateo College; Pat Keats from the Society of California Pioneers; Tom Carey from the San Francisco Main Library History Room; Susan Mitchem and Scott Bedio of The Salvation Army National Archives; Sue Schumann Warner from New Frontier; Jennifer Byrd, Karla Zens, Tina Schmitz and the Golden State Division Development Department; long-time Salvation Army historian Frances Dingman; Misty Jesse and The Salvation Army's Elftman Memorial Library which maintains an extensive theological and general education collection of 40,000 volumes; Colonel George Church and The Salvation Army Museum of the West directed by Captains Linda and Kevin Jackson, who—with a committee of representatives from throughout the Western Territory—are in the process of securing volunteer docents to help catalogue and preserve artifacts of Salvation Army history.

And finally, with great respect and enormous patience, Roger Rapoport and Richard Harris from RDR Books have brought this book to completion.

Thank you, gentlemen. Thank you all!

*As the story goes, when young Edith Smeeton sat on the general's lap,
her hair ribbon (barely seen here) tickled his nose. (page 77)*

INTRODUCTION

Telling the Army Story

Anecdotes and Memories

I never see a panhandler without remembering Tallulah Bankhead dropping a $100 bill into a Salvation Army lassie's tambourine and murmuring in her whiskey tenor, "There you are, dahling—I understand it's been a DREADFUL year for Spanish dancers."

—Herb Caen

It's always easy to start a story about The Salvation Army with an item by the late *San Francisco Chronicle* columnist Herb Caen. He often referred to the Army—not only in irreverent memories like this one from Tallulah, but regular pieces about unusual donors, unusual services and generally unknown facts.[1]

Such is the stuff this book is made of—anecdotes and memories that wouldn't ordinarily get into a history book.

It's not a definitive history. Not every fact is here. Not every name. There are other books with far more information about the Army than this and I encourage you to read them.

I'm not an historian. Not a theologian or sociologist. Not even a Salvationist. I simply want to share some of the Army's spirit . . . and to put into one book some of its favorite stories.

Salvationists aren't much on talking about themselves. They're a very private people, more used to focusing on clients and business matters than their own biographies. But they're proud of their history, even the days when people were throwing rotten tomatoes at them.

They love to tell stories about their beginnings. (William Booth, for instance, improved working conditions and helped stamp out "Phossy Jaw" face cancer with the world's first safety matches in a phosphorous-free factory.) They're excited when suddenly new photographs and newspaper clippings about a World War I Doughnut Girl are discovered in someone's attic. They're a close-knit family. They're passionate about their work.

Most people know about that work at Christmas. They know the Army kettles. They know the bells. They know the bands. But most don't know what the Army does the day after Christmas. They don't know the story of General William Booth and how the Army began in England at the same time the Civil War was ending in the United States.

They probably don't know the Army now invests as much as $18 million a year into San Francisco services.

Typically, Salvationists are fiscally prudent and hard working. They're do-ers. As a whole, they're basically good people quietly trying to do good things. But that's a blanket statement without real dimension. It hardly gives the whole picture.

Guys and Dolls is fiction. This is a story about real people.

Their lives are full of camaraderie, laughter, joy, frailty and pain. They see humanity at its strongest, and its weakest.

There are those who say the Army—like Franciscans and Mother Teresa—is a symbol of working with the poorest of the poor.

At its very best, in over a hundred countries, that work is a "partnership with the poor"—not hands down, but eye to eye. In America, it's often achieved through residential programs. In the dreary days of the past it was usually in soup kitchens and Skid Row dives. Today, it's often with low-income people who get by, just barely, from month to month.

In 1919, Bruce Barton from Batton, Barton, Durstine and Osborne Advertising Agency in New York coined the phrase, "A man may be down, but he's not out." It was done for The Salvation Army's Home Service Fund Raising Appeal and it touched a nerve everyone could relate to.

World War I was a big breakthrough for Army outreach. Salvation Army lassies gave away doughnuts and are still getting remarkably good press on that simple act. Veterans never forgot it.

These pages include memories of those doughnuts along with later disaster relief stories about the Army's relief work following San Francisco's big earthquake in 1906, Mexico City in 1985, El Salvador in 1986, San Francisco in 1989 and New York City after September 11, 2001.

Where did the phrase, "falling off the wagon" come from? Who knows the words to "Rum, by Gum?" How did the Army begin? How has it changed with changing times? How does an organization with a "meat, mashed potatoes and gravy" reputation work in a city that loves quiche?

The images here are varied—some of service, some of popular perception, some for reference. *Guys and Dolls,* Damon Runyon, Vachel Lindsay, George Bernard Shaw, John Phillip Sousa and St. Francis of Assisi are accounted for, along with *tzedakah* boxes, The Salvation Army's first kettle, the first brass band and the first bell.

Popular Culture

When Mae West offered the opportunity to "come 'n see me some time" she said it to Cary Grant playing an undercover cop dressed in a Salvation Army uniform.

That's Hollywood fiction.

Grammy and Emmy-nominated Greg Adams, formerly lead trumpet in the Bay Area funk band, Tower of Power, is the son of Salvation Army missionaries who served in China as well as San Francisco.

It's a fact most fans are not aware of.

When Jean Simmons played the role of Sister Sarah Brown in *Guys and Dolls,* her character was modeled after Captain Rheba Crawford, a young Salvation Army lass who led midnight meetings in the streets of the New York's Bowery during the Twenties.

That's a generally accepted truth.

Mark Twain said he put a cat, dog, rabbit, fox, goose, squirrel, doves and finally a monkey in a cage where they lived together in peace, even affectionately. In another cage, said Twain, he put an Irish Catholic from Tipperary; a Scotch Presbyterian from Aberdeen; a Turk from Constantinople; a Greek Christian from Crete; an Armenian; a Methodist from the wilds of Arkansas; a Buddhist from China; a Brahman from Benares. And finally, a Salvation Army Colonel from Wapping. When he came back two days later, they had beaten themselves into a frenzy. Nothing was left but gory odds and ends of turbans and fezzes and plaids and bones and flesh. The lower animals, Twain mused, got along just fine. It was the reasoning animals that couldn't seem to agree. In the end, they had "disagreed on a theological detail and carried the matter to a Higher Court."[2]

Anecdotes like these—short and reader-friendly—are included with history throughout this book. Drawn from many original sources and interviews, they illuminate the Army's raucous past and footlight its history. Some details have been gleaned from Salvation Army archives, family memorabilia and oral history passed down from generation to generation. But most of this book is a fresh look at a story vital to the heart of San Francisco and the Army's work from The Tenderloin to the Sunset District. Each section moves forward chronologically, offering a timeline on the Army's humanitarian work.

Here are some samples of what you can expect:

Beginning at the Beginning

The Army story began, as such stories tend to do, with the vision of one man, William Booth—a single-minded, peripatetic evangelist who had the theatrical instincts to mesmerize a crowd indoors or out and the promotional skills to make his voice heard throughout the streets of Dickensian London and eventually the world.

He also had a wife—the socially conscious, utterly pragmatic, wholly exceptional Catherine, who was one of the very earliest leaders in the movement to give women equal status in the pulpit.

For thirteen years they preached the gospel of salvation and then one day, with the stroke of a pen, they became an Army of Salvation.

How the Army Got Its Name

It was one of those moments when an impetuous action suddenly rings true and changes things forever. William Booth had started his work in London in 1865 under the name "New Christian Mission." In 1878, he and his son Bramwell were designing a flyer for the annual meeting.

It needed an attention-grabbing headline, thought William. "Volunteer Army" was his choice. But it wasn't Bramwell's. The son was already a full-time soldier and business manager in this operation. There was no way he was going to be called a volunteer. According to the history books, he stood his ground, protesting, "I'm a regular in this army, or nothing!"

Booth took a moment to reconsider and changed the copy to read "Salvation Army." It's been that way ever since.

Members of the Army are called Salvationists. What once was a very small mission has become the largest social service organization in the United States.

Volunteers, God Love 'Em

Volunteers, of course, are a key part of the story and always have been.

The late Fairmont Hotel owner Ben Swig understood the partnership in the best possible sense. After several hours of serving a Salvation Army Christmas dinner to the homeless in San Francisco, he announced, "Why don't you Christians go home now? We Jews can take care of things from here. This is your holiday. It's time for you to celebrate with your families."

For years, The Salvation Army family has included hundreds of volunteers on Celebrity Bell Ringing Day. In San Francisco's Union Square, Chief of Protocol Charlotte Shultz and friends merrily chase cable cars down Powell Street in search of tourist donations. On at least one occasion, her successor chair Terry Lowry has welcomed media pals in near torrential rains.

Bedrock support for the Army's work, of course, is in the advisory board and advisory councils. These are volunteers who come out for 7:30 A.M. breakfast meetings to give good ideas, counsel and feedback as full partners in this venture. They're friends of the Army in the best possible sense.

Queen Elizabeth and The Salvation Army's Chevy

A story about former divisional leaders, Lt. Colonels Ray and Bunty Robinson, will bring a smile to those who knew them during their tenure in San Francisco.

When Queen Elizabeth II arrived in port in 1983, they received an invitation to a reception aboard Her Majesty's Ship, the royal yacht *Britannia*.

Preparing for the event, Bunty Robinson found her husband busy waxing the family car. Soft spoken and basically very shy, she wondered aloud if perhaps it might be appropriate for guests to arrive in limousines for so grand an affair.

Robinson was a frugal man. Everyone knew that. He was British and certainly excited to see the queen, but not so much that he would ever consider putting on airs. He looked up for just a moment, surprised at the suggestion, and kept on waxing, never missing a beat. "No," he mused, "we'll take the old blue Chevy . . . and we'll drive it ourselves."

And so they did. The Chevy was parked near big black limousines. The reception was crowded with women in designer gowns. In deference to the queen, the Robinsons wore Salvation Army uniforms with stand-up collars no longer used in America at the time, but still worn in Britain, where the Army began.

Bunty wore a bonnet. A Scot by birth who had never had occasion to see the royals in Britain, she was thrilled by the occasion and framed the glove that shook the hand of the Queen.

She has it still.[3]

Ray Robinson was promoted to Glory in 1990, almost five years to the day after his adventurous trek with a 49-member volunteer relief team to assist victims of the 1985 Mexico City earthquake.

TELLING THE ARMY STORY

Promoted to Glory

Promoted to Glory . . . That's the Army's loving way of saying that after a lifetime of service, he died.

Early Salvationists had a decided penchant for giving colorful military descriptions to every aspect of their life. Some still remain. Describing a person's passage from life to death in military parlance, "promoted to Glory" says Salvationists not only believe in heaven, they believe it's something to be accepted and embraced, an assignment with eternal rewards.

In 2002 The Salvation Army in the Bay Area owned over 700 gravesites in Colma's Cypress Lawn Cemetery.[4]

Since 1924, the Army has held annual Memorial Day services there. In recent years, Salvationists throughout the country have filled the site with yellow chrysanthemums and American flags to decorate the graves.

Franciscans

In 1923, the poet Vachel Lindsay likened Salvationists to Franciscans of the Strict Observance in the earliest days of St. Francis. Salvationists, he said, were like the early mendicants, startling the church establishment by reaching out to the most degraded, most hurting souls in town—even after police, preachers and charity workers had given up.

Historian and California State Librarian Emeritus Dr. Kevin Starr was religion editor at the San Francisco Examiner in 1982 when he, too, compared founders of these two extremely pragmatic forms of spiritual outreach. He wrote that both St. Francis of Assisi and William Booth had a common belief:

"Christianity," he wrote, "should be vivid, immediate, joyous, practical, and animated by a special zeal for the poor and the downtrodden."[5]

In Context

For the purposes of this book, three dates should ring a bell. 1865—the year that William and Catherine Booth officially started their work in London. 1880—the year George Scott Railton and his seven "Hallelujah Lassies" officially arrived in New York to start the work in America. And 1883—the year Major Alfred Wells arrived in San Francisco to open the West Coast.

Through the earthquake of 1906, World Wars I and II, through the Depression and dot.com modern times, the Army continues its work 365 days a year, mostly in the South of Market area where it started.

Notes

[1] Some versions of this story say Bankhead left $50 instead of $100.

[2] "Letters from the Earth," from an English notebook in 1872.

[3] Interview with Mrs. Lt. Colonel Bunty Robinson, 2003

[4] In their book, "Pillars of the Past, At Rest at Cypress Lawn Memorial Park" Michael Svanevik and Shirley Burgett have produced a virtual Who's Who of community leaders buried there. Included, along with documented history, are profiles of Army officers Anna Allemann Butler who died in the 1906 earthquake and Japanese Divisional leader Masasuke Kobayashi. Their research was assisted by a book, "Comrades Promoted to Glory," compiled by Mrs. Colonel Henry Koerner (R) and Frances Dingman, Western Teritorial Headquarters Museum.

[5] Kevin Starr, San Francisco Examiner, Saturday, August 14, 1982

ONE

Bringing the Booth Legacy to California

General William Booth and the Bay Area Press

When he came to the San Francisco Bay Area in December, 1894, General William Booth did what any modern day celebrity would do.

He held a press conference.

It was his first trip to America after the publication of his book, "In Darkest England and the Way Out." His social service proposals were very controversial. People were curious. What would he be like?

A *San Francisco Chronicle* reporter waxed poetic the way journalists in those days were wont to do: "Not a drum was heard nor the beating of brass as the Napoleon of the religious world arrived. The founder and organizer of The Salvation Army, with 1,000,000 soldiers under his command came into town like an ordinary man, climbed into an ordinary hack bespattered with Oakland mud..."

Booth was on his way to a house on Piedmont Avenue in Oakland, which had been arranged for a meeting with "pressmen." That was his name for journalists.

Without hesitation, he immediately set the tone for the interview. "If you do not think my scheme is a fair and square one, then criticize; but if your conviction is the opposite, then give me a fair deal."[1]

The general held three public meetings in Oakland during his visit, a fact not lost on the folks in Sacramento. They had also wanted to see him and bemoaned his appearances in Oakland, which they called "over-cherished, over-pampered, over-lectured, over-entertained and over-newspapered."

The meetings at Mills Tabernacle were very dramatic, said the reviews, which were illustrated by line drawings. Although they were exaggerated caricatures, the artwork made it clear Booth was a commanding presence. He knew how to please a crowd.

The Booth Legacy Started with Soup, Soap and Salvation

Booth's success was a long time coming.

As a young man, he knew for certain he didn't want to be a pawnbroker, even though he had been an apprentice. He knew he'd never be satisfied as minister in an organized church. Eventually he left the Methodist Connexion to become an itinerant evangelist. But it wasn't until July 1865 that his ideals and reality began to converge.

Stunned, he stood in the streets of London's East End, appalled by the filth, degradation and utter poverty which the Industrial Revolution had left in its wake. Children begged for gin. Prostitutes beckoned from doorways. The stench was unbearable. People's lives in this part of the city were wretchedly poor. As he found himself in front of the Blind Beggar saloon surrounded by people who had little to live for, Booth began to preach a message which was seldom heard in this place... hope.

Talking to people in the slums rather than from a pulpit in a comfortable church sent his senses reeling. This was where he was needed. This was his mission. He could hardly contain himself. History books say he marched eight miles home to his wife, plopped down his big black hat and declared, "Kate, I've found my destiny!"

After that, life in the Booth household was never quite the same. With increasing zeal—in fact, night and day and evidently with little time for anything else—William and Catherine Booth began the work that would eventually be called The Salvation Army.

Breakfast Was Served

At first, the focus was salvation—not soup, and not soap. There were, after all, 500 charitable societies already doing business in East London. Booth was a preacher, not a social worker.

It was never his intention to start a religious movement. He encouraged people to go to established churches. But very proper Victorians didn't care much for riffraff joining them in the pews. Booth's converts didn't want to go to mainline churches and the churches, in fact, didn't much like the idea either.

So he preached with great fervor and enormous theatrical flair on street corners and then in tent meetings on a Quaker burial ground. And the longer he preached, the more he could see that people coming to the meetings had barely enough to eat. The next step seemed logical. He started serving breakfast. Critics said he was bribing people to come to religious meetings. Booth didn't see it that way at all.

"I am assured," he said, "that the soup and bread given here is all many poor creatures have to eat all the day through." By the time he had moved into the People's Hall, his soup kitchen was serving gallons of soup every day.

A Victorian "Power Couple"

It was a fortuitous marriage—William Booth, the visionary, and Catherine Booth, the pragmatist. Together, they gave birth to the movement management consultant Peter Drucker in the latter part of the twentieth century called the "most effective organization in the United States."

There was a fierceness to their commitment—he always thinking ahead to the next project and how to publicize it, she always prodding him on. Were his methods unorthodox? Certainly. Often they bordered on the outrageous. And the crowds loved it. Throngs of derelicts and impoverished people who would never have set foot into the very proper Church of England came to hear his preaching.

On the streets of London's seedy East End he founded an increasingly popular movement of the unwashed masses. To the establishment, he was an enigma. Who was this upstart preacher with untried ideas and utopian dreams? Victorian adjectives describe him best. He was notably dyspeptic, abstemious, irascible, autocratic. As his beard grew longer and whiter, he was the undisputed patriarch in his household and in his work.

Catherine, on the other hand, was the better student and ultimately the more persuasive and sought-after speaker and fund-raiser. It was she who served as emissary to socially upscale Londoners on the other side of town. It was she who nurtured the social services ministry and she who fought for women's equal right to participate. It was she who held the family together.

William was the General. Originally that was short for General Superintendent. After 1878, when the organization adopted its military identity, it was an easy step to make the transition on an advertising poster. If this was an Army, the top brass needed a military title. Catherine never had a title of her own. It was a sign of the times. Even though she was a recognized leader in her own right, in the best Victorian sense she was the woman behind her man.

There was a sweetness in their marriage, however, and a seemingly boundless energy in their work. On the edge of creative change, they were the ultimate power couple.

They had eight children—Bramwell, Ballington, Catherine Booth-Clibborn, Emma Booth-Tucker, Herbert, Marian, Evangeline and Lucy Booth-Hellberg—most of whom played a major role in The Salvation Army's development. Ahead of their time, Booth's married daughters assumed hyphenated last names.

The tenderness Catherine and William felt for each other was evident throughout their long courtship, their long love letters to each other when they were often far apart and, finally, after 35 years of marriage—when she learned she had breast cancer and perhaps only two more years to live.

On hearing the news, Catherine's initial sadness seemed not for herself, but the thought of leaving her husband alone.

"Do you know what was my first thought?" she asked

General William Booth

him gently as she told him the news. "That I should not be there to nurse you at your last hour."

Catherine was promoted to Glory on October 4, 1890. William lived 22 years longer.

In Darkest England

Catherine Booth was dying at the same time the ideas of William Booth, Commissioner Frank Smith and editor W. T. Stead were coalescing into the book, "In Darkest England and the Way Out." She refused surgery and refused morphine. But even in her sickbed, she was a major force in developing the book.

Booth and his followers had experimented with a variety of ministry approaches in the beginning years. By 1890, a social imperative had emerged. Booth "was settling on the fact that for the sacred to penetrate the secular, his perspective must be holistic."[2]

Characteristically, he used a very graphic illustration to make his point about helping street people.

He started with a horse.

Dead in a Month

Booth contended the people he was speaking of would be dead in a month without help. His arithmetic brought him to the conclusion that a full tenth of London's population were in this condition. A submerged tenth.[4]

He knew he couldn't tackle the whole of London's problems. If he could make a dent in just one level of a very stratified society, that was his goal. How to achieve it? He saw three solutions—a City Colony based on shelter, food and jobs, a Farm Colony outside the city limits where people could thrive, and finally an Overseas Colony where they could start entire new communities.

It was a plan based on food, shelter, jobs and fresh air.

The City Colony became the foundation of most of what the Army does in inner cities today. In America, Farm Colonies experienced a few years of limited success in Colorado, California (Fort Romie near Soledad) and Ohio. Overseas Colonies never really caught on.

"In Darkest England" was an immediate best seller. The first edition run was 10,000. It sold out the first day. A month later, there was a second edition of 40,000. It was gone at once, with another 40,000 already coming off the press. A year later, the Army's *War Cry* publication advertised a Fifth Edition of 200,000 more. All profits went to development of the plan.

The Cab Horse Charter

"When in the streets of London a Cab Horse, weary or careless or stupid, trips and falls and lies stretched out in the midst of the traffic, there is no question of debating how he came to stumble before we try to get him on his legs again.

"The Cab Horse is a very real illustration of poor broken-down humanity; he usually falls down because of overwork and underfeeding. If you put him on his feet without altering his conditions, it would only be to give him another dose of agony; but first of all you'll have to pick him up again.

"It may have been through overwork or underfeeding, or it may have been all his own fault that he has broken his knees and smashed the shafts, but that does not matter. If not for his own sake, then merely in order to prevent an obstruction of the traffic, all attention is concentrated upon the question of how we are to get him on his legs again. The load is taken off, the harness is unbuckled, or, if need be, cut, and everything is done to help him up. Then he is put in the shafts again and once more restored to his regular round of work.

"That is the first point. The second is that every Cab Horse in London has three things; a shelter for the night, food for its stomach, and work allotted to it by which it can earn its corn.

"These are the two points of the Cab Horse's Charter. When he is down he is helped up, and while he lives he has food, shelter and work. That, although a humble standard, is at present absolutely unattainable by millions—literally by millions—of our fellow-men and women in this country."[3]

Do You Like The Salvation Army?

This isn't to say everybody loved the Army's radical new proposals. Many didn't. T.H. Huxley called them "socialism in disguise" and spearheaded a running dialogue against them in *The London Times*.

To some, including the Anglican Church, Booth's ideas were considered quite outrageous and impractical.

People like "The Hound of Heaven" poet Francis Thompson thought the plan might fail. It would be a leap in the dark. But would it be such a bad thing for a community to make such a leap?" Some, he said, "make them often for far less clamorous cause."

The Venerable Archdeacon F. W. Farrar, D.D, wrote in the December *Daily Graphic* that Salvationists were "working heart and soul while many others contented themselves with idle waiting and barren talk . . . "

And perhaps, best of all, there was the ingenuous, no-punches-pulled response from a bishop, presumably a pillar of the establishment church, who was asked, "Do you like The Salvation Army?"

"Truthfully, I cannot say that I do," he replied, no doubt scratching his head at this upstart organization built on such unorthodox methods. "But to be honest I must confess I believe God does."

While Women Weep

While women weep as they do now, I'll fight;
While little children go hungry, as they do now, I'll fight;
While men go to prison, in and out, in and out, as they do now, I'll fight.
While there is a poor lost girl upon the streets;
While there remains one dark soul without the light of God,
I'll fight—
I'll fight to the very end![5]

—*General William Booth*

Catherine's Legacy Of Women's Rights

No one—in this century or the last—would ever have called Catherine Booth, wife of Salvation Army founder William Booth, an "uppity woman."

She most certainly was not.

But in 1860, even as she finally got up the nerve to march to the pulpit to give her first speech in public, she understood that women could and should be heard. She was a strong woman working hand in hand with a strong man. And when Paul's letter to the Corinthians was brought out to remind her that women were to be seen and not heard in church, her response was decisive:

"This is not a church . . . and I am not a Corinthian!"

Catherine took an early and important stand for a woman's right to preach. Along with Quaker women who were already well-established ministers, she and a very few others stepped up to the pulpit in the 1800s. An African–American holiness preacher, Zilpha Elaw, delivered her first sermon in 1819. Jarena Lee preached 178 sermons and wrote the first personal narrative by an African-American woman in 1827. Antoinette Brown was ordained by the Congregationalist Church in 1853.[6]

Six years later, in 1859, when Catherine was thirty years old and confident in her own opinions, she also was ready to publish her views.

She thought it was only a matter of time until women could preach. Common sense and public opinion would prevail. "Then," she said, "the doctor of divinity who shall teach that Paul commands woman to be silent when God's Spirit urges her to speak, will be regarded an astronomer who should teach that the sun is the earth's satellite.[7]

William Booth heard his wife's words and acted on them. Even in the earliest days, The Army's official rules and regulations reflected her thoughts.

"Women shall have the right to an equal share with men in the work of publishing salvation.
"A Woman may hold any position of power and authority within the Army.
"A Woman is not to be kept back from any position of power or influence on account of her sex.
"Women must be treated as equal with men in all intellectual and social relationships of life."[8]

Young women were the backbone of the early Salvation Army movement. Many came from the streets. Many were dirt poor. Still others were teenagers from the middle class and—thanks to George Bernard Shaw for reminding us—others like the fictional "Major Barbara" even came from the upper classes.

The uniform made them equal.

As the work progressed and the women's suffrage movement grew, many saw the Army as a place to exert their independence and get away from the restrictions Victorian life placed on them.

"The Greatest Man I Ever Met Was a Woman"

Evangeline Cory Booth was born on Christmas Day in 1865, the same year The Salvation Army was born. Her mother Catherine had been deeply moved by the story of "Uncle Tom's Cabin" and her first thought was to name her daughter after Little Eva. Ultimately, she chose Evangeline, saying it reminded her of the angel sent to announce Christ's birth. In the end, William registered the baby as Eveline (sometimes spelled Evelyne or Evaline).

Evangeline was headstrong. As an adult, she went back to her mother's choice, thinking it had a more dramatic flair. In fact, Frances Elizabeth Willard, founder of the Women's Christian Temperance Movement, suggested she use her full name—Evangeline Cory Booth. It lent dignity to a "woman with public responsibility."

As chief executive officer, Evangeline turned out to be a formidable figure. She was the Army's National Commander in Canada for eight years (1896 to 1904), National Commander in the United States for thirty years (1904 to 1934) and General of the international Salvation Army for five years (1934 to 1939).

Thomas Marshall, Vice President under President Woodrow Wilson, had some particularly choice words for her:

"The greatest man I have ever met was a woman. If, instead of being a Salvation Army lassie, she had been an actress, no one would have graced the stage as she would; and if, instead of being a Salvation Army lassie she had been a politician, it would not have been worthwhile for anyone else to run on the other ticket."

Changing Roles for Salvation Army Women . . . It Takes All Kinds

Even though the Army has always officially maintained the rights of women to serve in any capacity, it's pretty much like any other organization affected by the changing social fabric. Times change. Women change.

Historically, women in the organization have run the gamut from sweet-as-apple pie Doughnut Girls to USO volunteers and Rosie the Riveter relief workers.

Interestingly, no matter how far up the ladder they may go, a significant number of them can still tie a pretty bow.

As ordained ministers, they can marry and bury, daily provide a shoulder for people to cry on, organize a food line or clothing distribution in the morning, and still take time out to attend a Soroptimist meeting and be back in time to coordinate a kids' after-school program. In some appointments, especially small rural communities without staff, they're called upon to be everything from minister to bookkeeper to chief cook and bottle washer.

"I love the Army life. It was one of the pacesetters in Women's Lib, you know. We've always had equal responsibility, opportunity and rank. It has my total commitment and I'll remain 'til death do us part."
—*Lt. Colonel Pauline Eberhart quoted in The San Francisco Examiner, March 29, 1972.*

The Three Catherines

General John Gowans (R) tells the story of the three great Catherines in Salvation Army history: 1) Catherine Booth, feisty and influential wife of the founder; 2) their first daughter Catherine, known as "La Marechale," who planted the Army flag in France and Switzerland; and finally, 3) granddaughter Commissioner Catherine Bramwell Booth, who in the last years before her death at age 104, became a popular "chat show" guest on the BBC.

After retirement she went through a period of depression, feeling no longer useful. So, at 90, she started a new career as Army spokesperson.

Although the first Catherine's language was Victorian, said Gowans at a Salvation Army Heritage Conference in 2003, she presented a "bang-on, up-to-date image. She was vivid in her image and courageous in her substance. If we are a temperance church, it is because of Catherine. She decided and then talked the old man into it."

Would she have been disappointed in how long it's taken for women to achieve administrative appointments of significance in The Salvation Army? What would she say to young women going into the work today?

Catherine's admonition to those who followed her was blunt. She worried that women wouldn't reach their potential. Gowans called upon people training to become Salvation Army officers to study her writings on the library shelves of the Army's Crestmont College in southern California. "Don't expect a pleasant afternoon of it," he warned. "She'll make you think!"

Think pink. That may be the way some will remember the second Catherine, an electrifying leader who after a full life of service and ten children, as if to exert her independence, defiantly said, "I've worn navy blue all my life. Now, if I want, I shall wear pink."

In the audience listening to the General were retired officers and women who had given decades of their lives to The Salvation Army seated next to cadets just beginning the work. What was the impact of the earlier women's lives on the newest soldiers? One cadet's response was ingenuous. "The General has given women permission to be strong. I'm going for it!"

Today, at all levels, a number of assignments are increasingly specialized. Women are also taking on jobs ranging from the legal department to training college instruction, program management and business administration. General Eva Burrows, an educator in Africa for 15 years, was in charge of the international Army from 1986 to 1993. Women are now Territorial and Divisional Leaders. Lt. Colonel Bettie Love served as Administrator for the City of San Francisco until her retirement in 2004.

Others prefer traditional assignments—those that allow them to stay close to the community on a very personal level. Skills may vary. But the role of helper is probably what inspired them to come into The Salvation Army in the first place. For those who are married, having desks just down the hall from each other, participating in a joint ministry, is exactly the kind of job description many married Salvation Army officers cherish.

Fortunately, that's exactly how the system usually works. Husbands and wives have traditionally held parallel assignments. It's purely a matter of logistics. It's easier to move them around that way.

Ironically, for an organization that was ahead of its time in giving women equal rights, it took quite a few years to give them equal titles. There was a time when a new wife assumed her husband's rank, even if it was lower than hers, but that's changed now. Women who used to be called Mrs. Colonel are now called Colonel.

Can an officer be on call for the world 24 hours a day and still give her personal life the attention it needs? It depends on the person.

Some are supercharged. They're the ones who go to church on Sunday, wash clothes on Monday and do outreach the rest of the week, perhaps while squeezing in PTA meetings and soccer matches before standing at the stove presumably belting out "I am Woman, Hear Me Roar . . ."

Salvation Army women are just like the rest of the world. They're sometimes torn and sometimes just plain tuckered out.

It seems safe to say many would agree with suffragist Elizabeth Cady Stanton, another woman with a mission and, incidentally, seven kids. "My whole soul is in my work," she said. "But my hands belong to my family."[9]

Like other career women—and men—they love their work. They love their families. It's a balancing act.

General Booth Entered Heaven With His Big Bass Drum

It was August 27, 1912. At the last moment, Britain's Queen Mary decided to attend the funeral of William Booth, founder of The Salvation Army. An estimated 35,000 to 40,000 people packed the hall. And because there was no other room, the Queen was put in an aisle seat next to a former prostitute.

That's the way Booth's biographer Richard Collier told the story in his 1965 book, *The General Next to God*. Booth had saved the prostitute's life and according to Collier she had placed three red carnations on Booth's casket which remained there throughout the service.

Booth's epitaph had special meaning for her, "He cared for the likes of us."[10]

Collier painted a graphic image—Queen and prostitute seated next to each other. At the time of his death, Booth was honored by heads of state and beloved by thousands whose lives had been touched by the movement he founded.

He had scandalized the established church. His tactics were outrageous. His presence was larger than life. His death was noted throughout the world.

In America, the poet Vachel Lindsay was moved to write a poetic requiem charged with dramatic imagery. In it, he envisioned multitudes of people whose lives had changed dramatically through Booth's preaching. First in tatters, degradation and poverty, then in white robes of glory, a cortege of beggars, prostitutes and lepers accompanied Booth into eternity.

"General William Booth Enters Heaven" was published in 1913 and the vividness of those images caught America's imagination.

Lindsay was brought up in Springfield, Illinois. It seems only natural that his picture of the Army would start on a familiar street corner. In those days important gatherings were always held in Courthouse Square. That's where Lindsay decided heaven would be. He peopled the square with ranks of the lost who Booth—although blind when he died—would know intimately as those he had fought for all his life.

He based the poem's cadence on the gospel music, "The Blood of the Lamb" by Elisha Hoffman.[11]

> "Booth led boldly with his big bass drum. . . !
> (Are you washed in the blood of the lamb?)"

Anyone who's ever heard those words in Lindsay's own voice will never forget them. His high falsetto seems to caress the poem. The cadence pulsates. The sound of each word stays with you long after the recording is over.[12]

It was his first published poem and it propelled him onto the literary scene. For the rest of his life women's clubs, Rotary and other social groups begged him to recite it over and over again. Each recitation was a *tour de force*, not so much a reading but an almost vaudevillian performance. Audiences loved it and his other popular favorite, "The Congo." But his later work never received the same acclaim.

Lindsay grew despondent.

Courthouse Square wasn't the only place he had found The Salvation Army. This was a wandering minstrel lured by the song of the open road. For several years, like a hobo looking for work, he had literally traded poetry for food.

He knew first hand what services the Army offered poor people. That's because sometimes he was poor, himself.

"I know The Salvation Army from the inside," he said.

General Booth had focused on what he called the pauperized and degraded submerged tenth of the population. Lindsay knew what that was like. For a while, he too was submerged. On at least four occasions he slept at a Salvation Army shelter.

Christianity with Its Sleeves Rolled Up

Years later, William L. Stidger, D.D. interviewed Lindsay. The poet told him, "I wrote that poem because I have always looked upon The Salvation Army as Christianity with its sleeves rolled up—Christianity down in the gutter, on the streets, in the factories, in dark alleys, in prisons, in saloons, around the world where the down-and-outs congregate. I have always been a mystic, and yet I have always had respect for those who put Christianity to work down where the people live."[13]

Edgar Lee Masters—Lindsay's contemporary and biographer—said the poem was better than anything else the poet had ever written. Lindsay demurred. But he said the most important thing to him was that if Salvationists liked it, if it had integrity, that was all that mattered.

"I don't care a hang what the critics say about it so long as the Army and General Booth's friends like and understand it...if they like it, it will last long after my little hour and day."[14]

Lindsay was a visionary, an idealist whose moment of fame came, shone brightly for several years and ultimately faded. His poems received mixed reviews throughout his life and also in the history books. He was a man tormented by many demons, a tragic figure whose life never quite lived up to his illusions.

On the night of December 5, 1931, he quietly drank a bottle of Lysol and died.[15]

The Images Remain

"So long as The Salvation Army likes it . . . the poem will last."

Lindsay's words touch a poignant note, because The Salvation Army very much does like this poem.

Two Salvationist songwriters wrote ten Salvation Army musicals in their early years and ultimately both became generals of the international Army. They used Lindsay's poem as the basis for their play, "The Blood of the Lamb." General John Gowans wrote the words and General John Larsson wrote the music. Ask Salvationists everywhere what part they played and many will remember being among the crowd following Booth into heaven.

If It Has Integrity

In 2004, the words and images of the Booth poem rang out again. When Old First Church held its first San Francisco Song Festival and invited local vocalists to perform new music based on some of America's favorite poems, they invited The Salvation Army, too.

The spirit of poets from other eras and other places filled the room—Langston Hughes, Henry Wadsworth Longfellow, Thomas Lux, William Carlos Williams, Miguel Hernandez, Sarah Teasdale, John Milton, Emily Dickinson, Vachel Lindsay.

The words were familiar. Only the music was new. A new generation of composers had written award-winning compositions to the words of traditional poems. The audience loved it.

After the intermission, composer Bruce Broughton came to the stage.

Broughton's professional credits include twenty-one Emmy nominations and nine Emmy awards for TV productions, more than any other composer of original music in Emmy history. His music for *Silverado* won him an Academy Award nomination. *Young Sherlock Holmes* was nominated for a Grammy.

Wherever his love of music has taken him, it started with his Salvation Army upbringing in San Francisco. Both parents and four grandparents were Salvation Army officers. His grandfather, Brigadier William Broughton, was Territorial Music Secretary. Tonight the grandson played his original composition based on "General William Booth Enters Heaven," the poem Vachel Lindsay wrote in 1912.

Broughton addressed the music with the same kind of lyrical, pulsating, mystical emotion Lindsay brought to his poetry. His hands moved up and down the keyboard in a melodic rush of notes. The feeling was intense, just like the poem.

Banjoes, bass drums, sweet flute music and then in grand crescendo, a chorus of all instruments including tambourines—that's what Lindsay's written instructions called for. The music was to be a "tune that is not a tune, but a speech, a refrain" like those used in Army outdoor meetings.

Broughton captured the cadence. Alone on the piano, he translated all those sounds and all that emotion into one presentation. Accompanied by an enormous musical crescendo, William Booth marched into eternity.

Charles Ives and a Full Orchestra

When Bruce Broughton performed his version of Lindsay's piece, he was alone on the piano. In 1914, when the composer Charles Ives wrote his version, he used a full complement of orchestral sounds to illustrate the poem. A huge, larger-than-life rendition, his music was a rich, cacophonous blending of sounds. Craig Helms laughs as he remembers finding Lindsay's poem in a high school literature book and telling his teacher how he and his brothers got their first introduction to the poem and the music. While they were growing up, their father Gordon Helms would put on a 33⅓ vinyl recording of the music to wake the children on Sunday mornings! Mischievously, the boys would hide it, but eventually their dad always found it. It was a very effective alarm clock.

> **Already in India**
>
> The Army was moving in worldwide directions far faster than anyone could ever have imagined. In 1882 it started work in India. In 1883—the same year it came to San Francisco—it also started in Sri Lanka. By 1894 it was in Indonesia. When a tsunami hit those countries on December 26, 2004, the Army was already there with a network of 869 centers and 126 schools plus 27 hospitals and clinics. A total of 3,562 officers, cadets and employees were already working in the region. Issues of housing, water, sanitation, roads, jobs, health care and disaster preparedness were already being addressed. A 244-acre site in the Galle district has been allocated to The Salvation Army by the Sri Lankan government and will become an entire new community for 1,000 families.

Coming to California and Finding the Gold

Before anybody knew about the gold, San Francisco was a sleepy little town called Yerba Buena with about 400 people in it.

A year after the Gold Rush, it had grown to 40,000.

After Native Americans—who were here before anybody—came the Catholic Church and then a shipload of 238 Mormons led by Elder Sam Brannan.

Brannan was an enterprising fellow. When word got out about the gold, he was quick to get on the bandwagon and he did a land office business selling shovels in his hardware store. Gold miners rushed in. The Barbary Coast exploded, rowdy and disreputable. Next came wives and families. Next, the preachers.

Methodist, Presbyterian, Baptist, Episcopal, Congregational and Lutheran churches, Chinese temples and Jewish synagogues all were here. By 1850 there were at least 100 places of worship.[16]

The city grew dramatically. From its waterfront beginnings, the upwardly mobile middle class moved to the suburbs (in this case westward towards the ocean) and they took their places of worship with them.

For a while, Rincon Hill was fashionable. Union Square was a posh residential address with fine churches and synagogues. But San Franciscans who could afford it soon moved toward Van Ness Avenue and beyond. Behind them, they left the Barbary Coast waterfront, Chinatown and South of Market—which is where The Salvation Army came to stay.

William Booth, founder of The Salvation Army, was still a young itinerant preacher in England when the Forty-niners rushed to California for gold. During the last half of the 19th Century, he was busy establishing the Army on the other side of the world.

In fact, if it hadn't been for the Pacific Coast Holiness Association, the Army might never have gotten to San Francisco as early as it did. A copy of the Army's British *War Cry* magazine fell into their hands. The Army was making a dramatic impact in the working-class East End of London. The Holiness Association decided—with or without permission—to create a stateside version.

Resolutely, Reverend George Newton, a Methodist preacher, wrote to General William Booth, the Army's Founder. "We'll turn over our troops to you," he offered. It took Booth several months to honor the request.

It took 14 months before General Booth found young Alfred Wells and declared: "You are the man I've been looking for . . . Are you willing to go as a pioneer of the Army to California? They have been writing and rewriting for over a year for us to send someone to Frisco and the minute my eyes caught yours, something said, 'That is the man you are looking for . . .'"

Wells was ready! At 24, he was already a rising star. Booth had made him The Salvation Army's youngest major.

A battle had been going on in his heart about Foreign Service. He had an "innate desire to see more of the earth's wonders" so the trip to America satisfied his longing to see the world.

Wells relished new discoveries. On the way to California, he took a side trip to see the wonders of Niagara Falls. When his cross-country train made a rest stop in Winnemucca, Nevada, he walked half a mile to a nearby Indian reservation to meet the natives. He was the only passenger to go.

The First Meeting

Eager to begin the Army's official work in San Francisco, Wells held the first Salvation Army meeting on the West Coast at 809 Montgomery Street. (A plaque in front of the building marks the spot, although sometimes there's a potted plant or sandwich board in front of it.)

It was July 22, 1883, his second night in town. He was the first officer to travel alone from England to start the Army's work in a new country. And he spoke only one language, English. Yet his polyglot congregation that first night included Swedes, Norwegians, Danes, Scots, an Armenian and a Turk. By December 31, the growing Army moved to a larger building at 632 Commercial Street. From there, the little band of Salvationists was constantly on the road, reaching out to establish Salvation Army units in Oakland, San Jose, Stockton, Sacramento, Napa, Santa Cruz and Saucalito (sic).[17]

True to the Army's penchant for wading right into the middle of things, Wells headed straight for the Barbary Coast—where depravity reigned. Prostitution, debauchery, the infamous practice of Shanghai-ing sailors, drugs, gambling, thievery, murder, political corruption. You name it. The Barbary Coast had it all.

Was The Salvation Army up to the challenge? The *San Francisco Chronicle* took note of Wells' arrival and sent a warning to "the denizens in and about the Barbary Coast . . . for it is in this neighborhood. . .that the Irish major will establish his headquarters and sally forth daily and do battle with sin and the devil."[18]

Henry Stillwell

Captain Henry Stillwell had traveled from England to New York with Major Wells, but didn't come to San Francisco until seven weeks later. Once in California, he immediately started a Salvation Army outpost in San Jose—the third regiment.

Today, that's a fifty-mile cruise down the highway. Even in rush hour traffic, the commute takes only about an hour and a half. But in 1883, on horseback or by buggy, it was an overnight trip. During the first year, Stillwell and Wells saw each other only three times. Salvation Army activities were scattered widely, the work was hard and the troops were limited. Three years later when Wells and his wife left San Francisco, the Stillwells took their place.

The Stillwells were typical of early Army officers who were moved frequently to open the work in new territories. Career moves took the family across the United States. During their marriage, they were administrators of Divisions in Oregon/Washington; Illinois/Indiana; Minnesota/Northern Michigan; Kansas/Oklahoma; Missouri/Arkansas. They also served at The Salvation Army Farm Colony in Amity, Colorado, at National Headquarters and Men's Social Services for the Western Territory.[19]

The Indispensable John Milsaps

Eleven "well-inclined folks" formally enlisted in the first official Salvation Army in San Francisco. During the years following the Gold Rush, John Milsaps had been a wanderer looking for something, but not quite sure what. Without a plan and evidently without much luck, he drifted through the Dakota Territory, Wyoming, Colorado, Utah, Idaho, Oregon, California and Nevada—an itinerant postmaster, fruit farmer, mule herder, silver miner, railroader, carpenter who never settled in any one place for very long.

For him, The Salvation Army's arrival was a bonanza. Weary of being at loose ends, he became the first commissioned officer on the West Coast, traveled to Hawaii and the Philippines in the Army's name and became editor of the *Pacific Coast War Cry*. During the Spanish American War, he was the first Salvation Army chaplain officially recognized by the U. S. Army. Always a journalist, he kept his trusty camera with him.

Virtually every story about The Salvation Army's early days in the west include reference to "good old" Staff Captain and then Major Milsaps. There is a sense of the indispensable in the way he's described. Unmarried, he lived simply, gave any money he came by to soldiers poorer than he, seemed to have been embroiled in his share of brawls but, like other Salvationists, earned a reputation for never fighting back.

One Week End
IN SAN FRANCISCO
CHINESE GAMBLING HOUSE
TRADES UNION MARCH
JESUS CHRIST AND
MIDNIGHT MEETING IN THE DIVE
THEATRE
SUNDAY MORNING OPEN-AIR ON THE DOCK
MIDNIGHT OPEN AIR

Double Wedding on Fourth Street

Long before Starbuck's ever dreamed of serving café lattes, before Moscone Center, Yerba Buena Gardens, the Metreon and a City parking lot came to dominate the corner of Fourth and Mission, The Salvation Army was already part of the block.

In 1883 the address at 142½ Fourth Street was site of Alfred Wells' boarding house. In fact, if not officially, it was the Army's first headquarters in San Francisco.

Ironically, it's just a block and a half from The Salvation Army's current Golden State Divisional Headquarters at 832 Folsom.

Wells and Stillwell held a double wedding in this house. Their own. The two pioneers arrived in 1883. Their sweethearts—Captain Polly Medforth and Captain Mary Matthews (also called Polly by her family)—didn't come until a year later. After the long cross-country train ride that was more than a day late, the brides-to-be finally arrived, hot and dusty and understandably unhappy that no one was at the Ferry station to meet them. They walked the eight blocks to the boarding house.

Wells was too flustered to think about it at the time, but after the two brides and two grooms were married three days later, he chastised himself for asking a Methodist minister to perform the wedding ceremony instead of doing it himself.

Any one of the wedding party could have done the honors. All Salvation Army officers are ordained ministers.

Adding still further to the confusion, three weeks later Wells discovered that the marriage license he had secured in Oakland wasn't good in San Francisco, a different county. Sheepishly, he told Polly they needed another wedding.

Mary Stillwell

From the very beginning, women in The Salvation Army had the same opportunity for equal rank and equal hardships as men.

The Stillwells were appointed to open San Francisco Corps #2 on Market near Twelfth Street in April, 1885. There was a carpenter's shop on the ground floor with a large hall upstairs, a portion of which was made into their quarters where their first child was born. A soldier's life was transient and always sparse. Mary Stillwell once wrote that she and her husband lived in one room, "cooking in tomato cans on an oil stove and using the bed for a dining table."

Meetings in the Market Street building were usually crowded with rowdy young men intent on causing a ruckus. Previously, the hall had been used for the "cheapest kind of shows" so there was a history of unruly behavior.

The men wore their hats, smoked and broke most of the windows. They set fire to *War Cry* magazines, dressed up in women's clothes and caused such trouble that Mary Stillwell in exasperation once loudly called out, "Is there a gentleman in the house or are you all alike?" "I'm going to help that woman," said a man in the back. "Give me a dollar and I'll clear the whole crowd out." And that he did. Next day, the newspaper reported that prizefighter Patsy Lee had dispersed the crowd.

Mary was a trooper in every sense of the word. When the Stillwells were assigned to open a division in Portland, Oregon, she left to start the work four months early with a ten-month old baby in tow. At the time of Henry's death, she had eight children. The four youngest were ages 1, 3, 5 and 9. Just two weeks after her husband's funeral, she was selected to become Secretary of Women's Social Work for the Western Territory from Chicago to Honolulu, a position she held for the next 17 years.[20]

Fong Foo Sec

Intense poverty in China had forced thirteen-year-old Fong Foo Sec's parents to send him to America. Still floundering when he was seventeen, he joined The Salvation Army with a sudden sense of purpose and desire to succeed.

While working as a cook, he attended training classes at Territorial Headquarters, preached in the streets of several cities and—playing a bass drum—was a member of the evangelistic team on horseback that accompanied General Booth's visit in 1894. Short hand and typing were his night school studies. He became secretary to a major and finally, after eight years working with the Army in San Francisco, left to earn a series of advanced educational degrees (a BA, MA, PhD and LlD)—a rare honor for a Chinese in his days. Returning to China, he became a successful editor and volunteered for years in executive roles with the YMCA and Rotary.[21]

Captain Henry Stillwell, Major Alfred Wells and their wives, both known as Polly, (second row center) surrounded by their little band of soldiers. Photo courtesy of The Salvation Army Museum of the West.

Early Corps

Early Salvation Army units were known by their number, not their address, so it's sometimes difficult to follow locations of what was then a very transient Army. In the beginning, it wasn't Army practice to buy buildings. The organization was flexible, often on the move. When a storefront was no longer practical, Salvationists simply packed up and moved on.

Beginning in 1883, Major Wells, Captain Milsaps, Captain Alfred Harris and Lt. Ashley Pebbles were first leaders of Corps #1. Wells, Stillwell, Harris and Captain A. Reif were the first officers at #2, starting in 1885 when it opened as a Men's Training Garrison on Market Street and later reopened as a Women's Training Garrison in a building known as the Adelphi Theater on California Street. A year later it was the "Salvation Dive" on the corner of Pine and Kearny Streets. San Francisco #2 held the first midnight meetings in America.[22] (This eventually became the Harbor Light corps)

The first Chinese Salvation Army corps in the world (#3) opened in 1886. A *Chinese War Cry* was published in 1887. On October 1, 1890, Captain Sam Wood and Lt. Alex Foster opened #4, the Mission Corps.

Next came a series of corps opening in quick succession. Captain Kitty Wilson and Lt. Verna Riley opened #5 on August 2, 1891, and on the very next day, Captain Alice Smith (Bourne) and Lt. May Jackson opened #6.

In December, Captain Fannie B. Fields (West) and Lt. H. C. Anderson opened #7, the Fillmore Corps. Corps #8 opened on March 23, 1892, with Captain Ada White and Lt. Anna Alleman in charge.

The Henry Stillwell family, hard working pioneers in the early Army. Photo courtesy of The Salvation Army Museum of the West.

In May, 1892, Corps #9 opened with Captain Jennie Barnhart and Lt. Berriman at the helm. (According to a *Western War Cry* report, this was moved in 1925 to a "splendid" building purchased by Lt. Commissioner Adam Gifford at 76-78-80 Turk Street. It was "within one-half block of the famous Market Street, and one block from the same street going south . . . across from the property where the Army maintained one of its best corps nearly thirty five years ago." The new building was two stories high with a full basement and had two stores on the first floor (one was the Liberty Grill and Coffee Shop) and a "splendid" auditorium on the second.[23])

Corps #10 opened on January 13, 1893. Captain Thomson and Lt. G. W. Wade were in charge.

The Army Cruiser *Theodora*

Captain Joe McFee was an enterprising officer (cf. "The Lifeboat," page 69, and "The First Kettle," page 105). In need of funds, on September 5, 1893, he launched a fund-raising riverboat campaign from San Francisco to inland California. Using an engine that originally came from the *War Cry's* printing office, soldiers re-fitted a second-hand cruiser as a vapor launch.

The choppy San Pablo Bay crossing was "looking serious to those uninitiated," but they sang, "Oh, we never, never, never will give in, no we won't" and soldiered on through the smooth Straits of Carquinez to Crockett, Sacramento, Knight's Landing, Meridian, Colusa, Red Bluff, Butte City, Princeton, Sycamore, Grimes and Kirksville.

The trip was a major success—"hallelujah times" at camp meetings and eye-opening adventures for the crew. In nearly every town on the Sacramento River they secured agents to collect food and other materials for the Food and Shelter Depot and Rescue Home in San Francisco.

The early years were hard times for the Army. Numbers were low. If five soldiers turned out on a march, it was considered a success. Collections were erratic—sometimes as low as $16 a month. Rent was $25. But John Milsaps' diaries showed he could get by

BRINGING THE BOOTH LEGACY TO CALIFORNIA

on $1.30 a day by eating at Dennett's Restaurant at 749 Market Street in San Francisco. It was a temperance restaurant and the discounted prices were just right for a Salvationist pocketbook. That's where Milsaps met George Montgomery, a principal owner.

Millionaire Salvationist

George Montgomery was a stockbroker at 324½ Montgomery and had a seat on the Pacific Stock Exchange. He raised a fortune and lost $60,000 in one day, considered suicide, and went from being listed in the San Francisco Blue Book social register, living at the Palace Hotel and being a member of the Bohemian Club to joining The Salvation Army.

After his initial business failures, illness, conversion and marriage to faith healer Carrie Judd, Montgomery and his wife started The People's Mission at Pacific and Sansome.

On August 2, 1891, they officially turned it over to The Salvation Army and it became San Francisco Corps #5. On Thanksgiving the Montgomerys signed on as honorary Salvation Army officers.

As committed as he was to missionary service, however, Montgomery wasn't quite ready to abandon his entrepreneurial instincts. He saw his talent for making money as a way to support good works, so he and other Christian businessmen like Alfred Dennett and Charles Crittenton (who established Florence Crittenton Homes for pregnant women) met regularly to determine how best to invest in the community. A couple of questionable schemes drew public attention and people began to question the propriety of Christian participation in speculative business investments. In 1900, some Salvation Army soldiers across the country had purchased stocks in one of Montgomery's ventures. When the Army's National Commander found out, they were reprimanded and quickly transferred. Montgomery, himself, had second thoughts and in his diaries resolved to scrutinize his business ethics more carefully. Poor Mr. Dennett, owner of the restaurant, eventually filed for bankruptcy.

From 1882-1885 Montgomery gave the Army parts of his 67-acre Beulah tract of land in the Oakland foothills to build a Rescue Home, a Home of Rest and an orphanage. In 1904, he sold the Lytton Springs resort in Sonoma to the Army, which turned it into a home for orphaned girls and boys and today still uses it as a working farm for alcoholics in recovery.

Montgomery would have liked that. He was a strict teetotaler.[24]

Joe the Turk—Free Speech Activist

Anyone with a love for San Francisco's more memorable characters knows about "Emperor Norton, Protector of Mexico." In 2004 the Bay Bridge was nearly renamed in his honor.

Staff-Captain Joseph "Joe the Turk" Garabedian

But have you heard about The Salvation Army's Joe the Turk (1860-1937)? He was a character, too, and a fiery free speech activist.

Joseph Garabedian was one of the most colorful of the Army's earliest officers in San Francisco, and he played an important part in establishing the legal right of Salvationists to parade, hold open-air meetings and play a horn in a public place. He had a clear understanding of how to attract attention and made the most of it, believing—usually correctly—that if he was arrested for preaching on the streets, he could demand a jury trial and win. Given the choice of siding with rowdies or The Salvation Army, respectable church-going jurors usually took the side of the Army teetotalers.

People called him the Turk, but Joe was actually Armenian, a six-foot tall bruiser in crimson pantaloons and jacket, with an outlandishly decorated umbrella and Turkish fez. He looked like a prizefighter and was just as pugnacious.

Did they send him to jail? It was just another opportunity to preach. Sometimes it was through the windows to the crowds outside. Often it was to his cellmates. They say he once converted 20 prisoners in a single night.

In cities around the country, Joe whipped out his indelible pencil and once even red and black paint to embellish cell walls with signs saying, "Jesus is the drunkard's friend." When a judge rebuked him, Joe leaned forward with a friendly word of advice: "You're always drunk, judge. Jesus is the friend you need." In MacComb, Illinois, he literally ran a crooked mayor out of town and took over his job for six weeks.

Staying overnight in a "widow woman's" hotel, he stamped the pillowcases, wallpaper and linens with his favorite slogan "Jesus is mighty to save." As the story goes, she balked at first, knelt to pray, then asked him back to decorate her other rooms!

Before he became a kind of one-man-band for saving souls, Joe was already saving shoe soles in his shoe repair shop at 48 Sacramento Street on the Barbary Coast. The shop was next door to Finnegan's saloon. There, through a hole in the wall, he could access his regular quota of beer.

After he joined The Salvation Army, presumably he plugged that hole with an Army catalog, festooned the walls with colorful flags and decorated the soles of customer's shoes with S's made from brass rivets.

What did the S stand for? Take your pick. Joe said it was *Saved from Sin, Saved and Satisfied,* and *Salvation Soldier.*

It was also an inspired bit of marketing to promote his day job—*Solid Soles.*[26]

Notes

[1] *San Francisco Chronicle,* Sunday, December 16, 1894

[2] *William Booth, The Development of His Social Concern.* Monograph by Lt. Colonel Paul Bollwahn, ACSW, CSW, National Social Services Secretary for The Salvation Army, 2002

[3] "In Darkest England and the Way Out", pp 26-27

[4] Ibid, pp 30-31

[5] Reportedly from William Booth's Farewell Address at Royal Albert Hall in London, May 9, 1912

[6] Wesleyan/Holiness History of Ordaining Women, www.nazarenepastor.org

[7] "Female Ministry," a paper written by Catherine Booth in 1859

[8] Orders and Regulations

[9] "Not for Ourselves Alone, The Story of Elizabeth Cady Stanton and Susan B. Anthony" by Geoffrey Ward, 1999. Made into a documentary film for the Public Broadcasting System by Ken Burns and Paul Barnes.

[10] "The General Next to God," by Richard Collier, E. P. Dutton & Company, Inc. 1965. pp 247-48

[11] Gowans and Larsson Official Website, www.gowans-larsson.com

[12] In January, 1931, Lindsay recorded some of his poems. The discs were made by William Cabell Greer of Columbia University, and issued on 78 rpm records by the National Council of Teachers of English. Caedmon Publishers' long playing record "Vachel Lindsay Reading 'The Congo,' 'Chinese Nightingale' and Other Poems" made them available for the first time to the general public.

[13] "Christianity With Its Sleeves Rolled Up," *War Cry,* July 1, 1944.

[14] *Ibid.*

[15] Further insights into the life and work of Vachel Lindsay may be found in "Profile of Vachel Lindsay," compiled by John T. Flanagan, Charles E. Merrill Publishing Company, 1970; "Vachel Lindsay, Fieldworker for the American Dream" by Ann Massa, Indiana University Press, 1970; and Modern American Poetry, An Online Journal and Multimedia companion to the Anthology of Modern American Poetry (Oxford University Press, 2000) Edited by Cory Nelson, Department of English, University of Illinois at Urbana-Champaign.

[16] "Sacred Places of San Francisco" by Ruth Hendricks Willard and Carol Green Wilson, San Francisco Alumnae Panhellenic, Presidio Press, 1985

[17] "Beginning of The Salvation Army in America" by Alfred Wells, 1925

Notes, continued

[18] *San Francisco Chronicle*, July 1883

[19] Lt. Colonel Henry Stillwell obituary, *War Cry*, 1905.

[20] "My Life," speech given to Salvation Army cadets at Atlanta College by Lt. Colonel Mary Stillwell, January 3, 1946

[21] "For My Kinsmen's Sake" by Lt. Colonel Check-Hung Yee, Chapter 11

[22] "San Francisco # 2 Corps" by Sergeant-Major Arthur B. Barker. The Conqueror

[23] "A New Home for San Francisco No. 9 Corps" article with picture in *Western War Cry*, March 14, 1925

[24] "Businessman for the Gospel," *Assemblies of God Heritage*, Volumes No. 1 and 2, Spring and Summer, 1989.

[25] *Western War Cry*, March 1925.

[26] Joe the Turk information drawn from "A Look Behind the Legend," by Daniel A. Bazikian, Territorial Historical Commission Biennial Meeting, September 19, 1981. Materials collected by Lt. Colonel Houston Ellis (R): "The Officer," December, 1893 and June, 1894. "The General Next to God" by Richard Collier, 1965 – pp 170-173.

Early Press Clippings

"There is no false delicacy about the methods of The Salvation Army. They go down into the slums. They meet the poor and the destitute on their own ground. If those who are charitably inclined wish to have a feasible method of reaching the poor without any doubt of the money being honestly and intelligently disbursed, and not wasted on imposters, we believe they could not find a better almoner for their bounty than The Salvation Army."

—*Argonaut* editorial
quoted in Salvation Army publication,
"The Conqueror," January, 1894, p. 84

"Grotesque, but helpful; absurd, but a blessing to mankind; narrow in creed, but admirable in deed; an angel of good cheer in Harlequin's clothing; a message of peace. With a bass drum for an orchestra; uplifting fallen man, while the ghost of murdered music flees at its approach; ridiculously divine; preposterously glorious —This is The Salvation Army.

"The Salvation Army justifies its existence daily—red shirts, tin badges, bass drums and all. If you don't believe it, ask San Francisco's poor."

—*The Wave*, a San Francisco society journal
quoted in "The Conqueror," January 1894

Come Home

The Salvation Army's *War Cry* publication is widely distributed, making it a natural venue for Missing Person ads. These were printed at the turn of the twentieth century.

"Persons wishing to make application should address Colonel Wm. Evans, 1139 Market Street, San Francisco, Cal. and are requested to send 50 cents (or more, if able) with their letter to help defray the expenses of postage, etc. "Do not however, hold back your inquiry if you cannot afford the small amount."

Shields, John J., missing since 1880; was then in Fairplay, Colorado; supposed to have been seen in Modesto, Cal., 8 years ago; has light hair, inclined to be curly, and bluish gray eyes; height about 6 feet.

Opal, Edith, left home in Dec. 1894, presumably to get married to a man by the name of Waldon; supposed to have gone to Texas; she is now 31 years old; 5 feet, 4 inches tall; dark complected. Sister at Hartsells, Ala, wishes to hear from you.

Shearer, Fred Wallace, his mother wishes to know his whereabouts; last heard of in Purcell, Indian Territory, suffering from consumption. Write to brother, 137 Seventh Street, San Francisco, Cal.

Cunningham, Fred, on his deathbed wishes to hear from his mother, now Mrs. Paterson, or any of his brothers or sisters. His address is Mojave, Cal.

Bryant, Mrs. Robert, maiden name, Hannah Morris Bosworth; went from New Hampshire about 1853; has several children, some of whose names are Ella, Emma and Alice. Emma married a Mr. Graham.

Gill, Ben W., six feet tall; light complected; blue eyes; hard of hearing; a mining expert; last heard from at Grass Valley; was then planning to return to Placer County, where he had worked in a mine during the winters of 1896-1897. Mother very anxious. Will he please write her or persons knowing him, communicate with Mrs. Fannie N. Reno, Chico, Butte Co., Cal.

THE WAR CRY

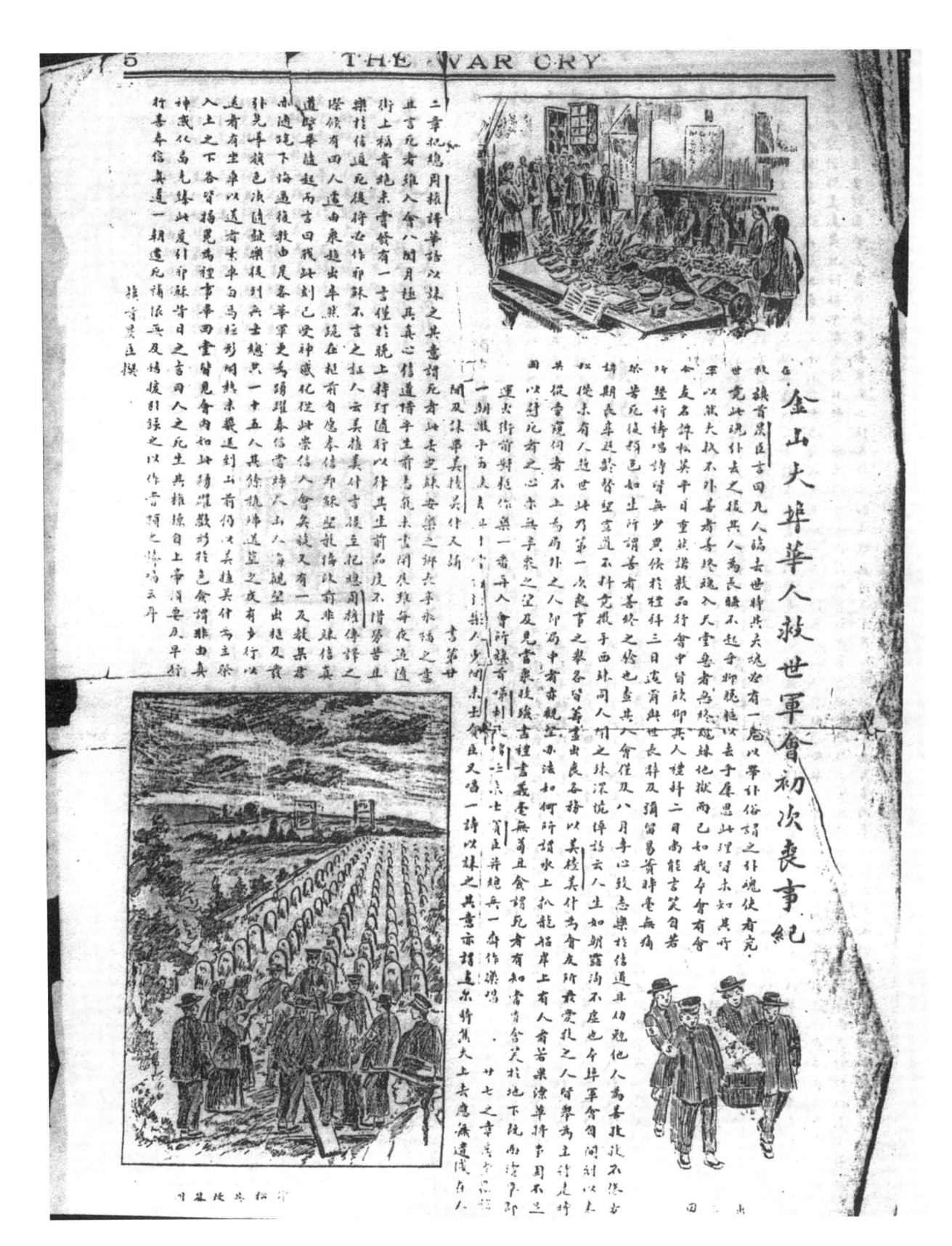

金山大埠華人救世軍會初次喪事紀

救鎮首影臣吉田几八病去世將共失逃必有一念以等俗韶之朴魂使有充
世充此現朴去之後共人為長驗不起手柳桃柩以去于庫思此理得末知其行
軍以散大拟不外善脊終魂入于天堂島為於砲朱地獄兩己如我今會有會
友名許松英平月重技諧教品行會中醫欲即其人健科二日尚能言笑自若
所整行詩唱詩唱時無少匙依於詩品十三日遠有興世長科及續留易貸時意無病
於苦厄後類包如生所謂善有善於之餘也盡共入會作及八月卒心致惹樂於信道
期長奉趣勸督望靈道不料克披于西朴司人們之珠深詭譎謫云人生如朝露淘不座也今年軍會旬間到以朱
被德朱有人逝世坦乃第一次良年之款各皆薷善出良各務以朱桂朱什為會友所装衆礼之人將聲為主詩是將
其從會覩死之不上為兩外之人即為中有奇觀怒不法知何行謂冰上扒龍船岸上有人看苦果涼華拌車圓不之
運次街前時起信因共一期散于而去亥十人人會所詠香佛制久廿上鋤忌又晤一詩以朱之其竟亦詩遣尔特將大上去應無遗後在人
聞及朱華美核朱作又新書第廿
一期散于而去亥十

二事記憶用椹譯華語以朱之其意詩華於八閒月極共真心信道傳年生肯為能朱畫間發程來信遊
街上稠音經尔賛有一言詩上拦訂適行以作共生前品及不情劳普且
樂作信道死後將必作卯珠不言悅在祖有自虞本信邪非朱信真
道聲華隨起而吉日我共刻己受神咸化祉此朱信人會英銳又有一及救異書
朱道號下悔通娓救由見各華軍更為首躍奉信書詩人山人每視堅出起及衆
排兒華散包次逗輆救到六士想兌一十五八共徐桃娷遘荐之故身步行以
送有有生牵以白鳥任形閗共朱嚴送到斉牟匹肉挂美什為
入土之下各皆非四呈肘見奇朱白挂肉食諧非由真
神威化鳥乞蜾此废引卯穌背月之吉因人之死生具推掾自上帝惜無及揚援引绿之以作音誦之蕨碼云丹
行善本信奉遣一朝遣卯穌諸依無及揚援引绿之以作音誦之蕨碼云丹

推官吳廷撰

The Chinese War Cry, *published in 1897 at 1139 Market Street.*
"Price 5 cents a copy or 75 cents per annum, postage paid, or $1 to foreign lands."

THE "CHARIOTEERS."

BY MAJOR KEPPEL.

We TAKE PLEASURE this week in presenting our readers with a picture of our "Charioteers", who will start on their trip to the mountains and mining camps, in the interior of the State, immediately after the close of our Camp-meeting—July 4th.

Although at this writing arrangements are not completed, yet in a most miraculous way God is enabling us to get things in shape, and showing His approval of our scheme already in that way.

We are aware that for the past two or three years, there has been some special campaign inaugurated for the summer months, which, we learn, were successful, and accomplished much good (especially the famous "Musical Cavalcade" of last year.) At the same time, this latter had its drawbacks and disadvantages, (acknowledged by the men themselves) which we will try to avoid this year.

One of the disadvantages was the "forced marches", leaving men and horses in a very worn condition—the men unable to keep up to go through with their meeting. The second disadvantathen. The following is, we think, a fair representation of the party and their plans.

We have secured a large stage coach weighing over 2,000 lbs., with a carrying capacity of nearly 7,500 lbs.; having seats for about twelve persons, with two large "boots" on back and front, into which baggage can be stowed. We propose to put six heavy, strong horses to this conveyance. On the rear end we will pack our large tent, capable of seating 800 persons. We will take the tent poles and other fittings along. The poles will be cut in two, with ferrules on, which will make them portable and convenient, without reducing their strength.

Now you can imagine them starting out, whip cracking, horses prancing, colors flying, stage wheels rumbling, heaven rejoicing and hell trembling. Off they go for a town about twenty-five miles away, where they will erect their tent, find a nice place for their horses, and for five or six days' go in for a proper waking up time, house to

each of the horses, two on the driver's seat, and the drums on top of the stage, which we think will prove quite an attraction every evening. We expect to have a first rate, regular stage driver to take charge of the horses, even if he has no other special ability. It will take a smart cool-headed man, we have no doubt, but God will supply our needs in this way I am sure. An artist will be another necessity of such a wonderful expedition, and already I have seen the "Warre Cryyee Mannee's" eyes sparkle at the thought of having much interesting manuscript, and many pictures illustrating and describing many wonderful scenes. Already Captain Sam. Wood has gone on horseback through the country, arranging for the campaign, securing locations for tent, lumber for seats, mapping out the roads, enlisting newspaper and Christian sympathy along the route, and in general preparing the way for the Lord's "Charioteers."

We are still in need of a few men such as described in my last notes; strong young men who can play some brass instrument

been fully restored through faith in the Great Physician. Praise His name.

We have already arranged with a Salvation soldier who for years was a stage driver, and unless something unexpected occurs he will probably be put in that important position, the only drawback is that he does not combine music with his muscularity, (he weighs over 200 lbs.)

We propose to muster the men on the Camp Ground for service, and practice, at "Trestle Glen" during our Camp Meeting from June 24th to July 4th, and where we expect a glorious time.

A special hat and summer jacket has been adopted resembling those worn by the Cavalcaders, with the addition of canvas leggings, which on the whole will give the boys a warlike appearance. A large supply of WAR CRY's, song books and Army literature will also form part of the outfit

Early Salvationists tamed six wild horses to pull their Charioteers on an 800-mile camp meeting trek through the Mother Lode.

BRINGING THE BOOTH LEGACY TO CALIFORNIA

Lost in America

The Salvation Army's Scandinavian Corps (#11) opened in 1899 and ultimately worked in three locations —the waterfront, 11th Street and 19th Street. Its original soldiers were seamen and their families who came to America in great numbers during the early 1900s. The Scandinavian Salvation Army flourished during those years in the Eastern, Central and Western Territories. Eventually, as immigrants settled into the American scene, old timers integrated into English-speaking corps and there were fewer and fewer young people to take their place. The waterfront corps officially closed September 30, 1968.

An old *War Cry* tells the story of a man in rags, disheveled and unkempt, who wandered into the waterfront Salvation Army meeting. He was stunned to see his picture on a wanted poster.

"Why am I wanted?" he asked. And he was amazed to hear the answer. He was heir to the $15,000,000 estate of his "immensely wealthy" parents in Sweden. Thousands of dollars had been spent trying to find him! When he stumbled into the Army, he was suffering from amnesia.[25]

Lt. Colonel Harold Madsen and Scandinavian Divisional Officers at the entrance of the Swedish corps on 19th Street between Valencia and Mission (Date unknown). Photo courtesy of The Salvation Army Museum of the West.

26 THE BELLS OF SAN FRANCISCO

A Selected Family Album—
The Salvation Army in San Francisco

Lt. Colonel Victor Newbould with Rabbi Joseph Asher, the Very Reverend David Gillespie and Monsignor Robert F. Hayburn at Grace Cathedral for the Centennial Celebration of the Army's 1880 arrival in America.
Photo from The Salvation Army Northern California and Nevada Divisional Archives.

ABOVE: **When they were Majors**, Phyllis and Check Yee posed for this studio shot with children in their congregation. Today, several of these babies attend the Asian American Corps, a second generation of Chinese Salvationists in San Francisco. Photo by Ron Toy.

LEFT: **American-Born Chinese Salvationists** attend the Yerba Buena Asian American Corps started by Majors Ron and Keilah Toy. Photo by Russ Curtis.

28 THE BELLS OF SAN FRANCISCO

TOP: ***Christmas on the road.*** *In the early 1980s Lt. Colonel Victor Newbould and Majors George Church and Rob Saunders took Divisional Headquarters officer staff to the streets of San Francisco on a motorized cable car. Photo from The Salvation Army Northern California and Nevada Divisional Archives.*

CENTER: ***Major Jerry Gaines*** *with senior advocates at the Senior Meals Picnic. Photo by Kazuhiro Tsuruta.*

LEFT: ***Lt. Colonel Bruce Harvey*** *got an assist from comedian Michael Pritchard as he boarded a cherry picker to unveil the billboard announcing the Harbor House/Gateway project. Photo by Kazuhiro Tsuruta.*

BRINGING THE BOOTH LEGACY TO CALIFORNIA 29

ABOVE LEFT: **Major Grace Phillips.** Photo from The Salvation Army Northern California and Nevada Archives

ABOVE RIGHT: **Major Chris Buchanan with Laurie Williams** (Robin's Mom). Photo by Russ Curtis.

RIGHT: **Lt. Colonel Ardys Newbould and Mary Jane Lee** of the San Francisco Women's Auxiliary. Photo from The Salvation Army Northern California and Nevada Divisional Archives

THE BELLS OF SAN FRANCISCO

ABOVE: *Envoy Elsie Lantz, Majors Bob and Carol Pontsler, Lt. Colonels Richard and Bettie Love, Lt. Colonel Bunty Robinson; Top row: Envoy Gene Lantz, Ken Iremonger and Captain David Eng*

BELOW LEFT: *Captain Chris Kim. Photo by Russ Curtis.*

BELOW RIGHT: *Harbor Light Corps Sergeant Major Ernie Henderson. Photo by Russ Curtis*

BRINGING THE BOOTH LEGACY TO CALIFORNIA

ABOVE:: **Lt. Colonel Ray Robinson.**
Photo by Steve Ringman, San Francisco Chronicle

TWO

Disaster Relief

San Francisco in Ruins
Greatest Catastrophe of the Century

The twentieth century was barely six years old when this headline ran in The Salvation Army's national publication, *The War Cry*, dated May 5, 1906.

"The bell tolls for doomed San Francisco! Earth's forces have combined to wipe out a noble city from its face by means of Earthquake and Fire! Oh, the mutability—the transitoriness, the unimaginable littleness of things human when placed in contact with the mighty forces of Nature…"

All hyperbole aside, can you imagine the shockwaves felt by people on the other side of the country when they heard about San Francisco's 1906 earthquake?

Unlike the 24/7 news coverage of the Loma Prieta earthquake in 1989, Dan Rather wasn't around in 1906 to give a minute-by-minute account of aftershocks and fires. The known news was huge. Fear of the unknown fired their imaginations.

The basic facts of San Francisco's big quake of 1906 are these. The date: April 18, 1906. Time: 5:12:05 a.m. (That's when the San Francisco Ferry Building clock stopped.) Duration: 46 seconds.

Early estimates said 478 people died. After intensive research, San Francisco City and County Archivist Gladys Hansen now says these figures were probably closer to 3000.[1] At the time, scientists recorded earthquakes with a Rossi-Forel scale of intensity. On a scale of one to ten, this was a nine. Later, using the Richter scale, it was estimated at about 8.2.[2]

The epicenter was at Point Reyes and damage occurred all along the San Andreas Fault from Fort Bragg to San Juan Bautista. Santa Rosa was devastated, including The Salvation Army's corps building there and its Golden Gate Orphanage at Lytton Springs—but like Hollister, Watsonville and Santa Cruz in the Loma Prieta earthquake of 1989—everything that happened in those communities took a back seat to what was going on in the city by the Bay.

All eyes were on San Francisco.

City on Fire!

Suddenly, it not only was a city of shattered buildings, it was a city on fire.

A city of wooden houses.

Three days of fires followed the initial damage caused by the upheaval of the earth. Dozens of fires broke out, one after another, at a dizzying pace. Gas and water mains broke throughout the city and there was increasingly little firefighters could do to control the blazes. Ultimately, fire marshals made a major decision to create a firebreak by dynamiting the elegant houses on Van Ness Avenue. In the end, 490 city blocks were destroyed.

From the very beginning, The Salvation Army had concentrated most of its work in the South of Market. This was one of the first neighborhoods to go.

If you look at an early map of San Francisco you can see the original shoreline where the bay was later filled in. During the Gold Rush, thousands of Forty-Niners sailed into port so eager to get to the gold mines they simply left their boats to rot off shore. Later, the dunes south of Market Street were leveled and pushed into the Bay. On this "made ground" or fill, once part of the old Mission Bay Swamp, hundreds of wooden boarding houses had been constructed. During the quake, some of them around Sixth and Folsom literally sank two stories into what once were lagoons. The ground liquefied. Ramshackle wooden structures erupted in flames. People inside were trapped.

Streets were in chaos. In 96 hours Southern Pacific train and ferry service reportedly carried away 225,000 refugees.[3] As residents raced for safety, Bay Area Salvationists immediately went into action following the quake.

True to form and with its inimitable, almost uncanny knack for staying out of the media's way, The Salvation Army quickly deployed many of its forces to the other side of the bay, where already existing facilities could be used for relief activities. (All but two of its San Francisco buildings were destroyed.)

In Oakland, Salvation Army personnel met hordes of refugees arriving from the City by ferry. They set up feeding stations and turned Beulah Park into a tent city where refugees lived for weeks. Chinese Salvationists from San Francisco played a major role in feeding thousands. After the military took over some of the responsibility of running Beulah Park, Salvationists returned again to San Francisco to assist in continuing recovery efforts.

All this information trickled back east. As soon as she heard the news, The Salvation Army's national commander Evangeline Booth responded "like a calvary horse to action" and kicked off a fundraiser in New York's Union Square that ultimately raised $12,000. On Memorial Day of that year, she brought $15,000 to San Francisco for the relief effort and in a Golden Gate Park rally spoke to a mass meeting of refugees who had lost their homes.[4]

Commander Evangeline Booth Spelled It Out—Do Something!

Rallying the troops in New York, Evangeline Booth had done what she did best, rousing the crowds with the famous words her father William Booth had used when he had worried how to help homeless people under a bridge in London. It was a no-nonsense plea—do something!

"You know what war is!" she exhorted. ". . . every man down the line is pressed into service.

"Some lie in the trenches, some reconnoiter, hunting for the enemy, some stand by the staff. Some roll up bandages, some prepare the rations, some come along with the ambulance, some sit in tents over maps planning, some manipulate telegraph wires, some blow the bugle, some lift tents, some, standing by the wounded and dying, form the front line, where shot is the heaviest and peril the greatest. But all hands are to the rope, all men in the fray."

Her words rolled off the tongue in eloquent cadence.

"Put your hand to the rope, throw your practical sympathy into some offering, on to the great sea of assistance that is going out to 'Frisco. You may not be able to pay for a carload of provisions, or nurses, or doctors, but you can send a bandage, or blanket, or bread, or a tent.

"Do something—do the most you can—do it well—do it now."[5]

Indoors, at the Hippodrome, the commander produced an extravaganza of grand proportions to rally the troops and send her soldiers off to battle. Onstage, as she sang a song of love and told the story of a broken heart, 260 women in white arranged themselves in the form of a cross. Army men in red uniforms formed an outline around them and officers in dark uniforms formed yet another outer line. It was the Army's 26th anniversary congress.[6]

Mark Twain Said the Poor Man Gives the Most

Mark Twain also put in a good word. At the time of the earthquake, he hadn't been in San Francisco for 38 years. But like other expatriates, he considered himself a Californian no matter how far he was from the West. His logic was unquestioned. Once a Californian, always a Californian.

The April 22 *New York Times* covered a meeting of leading citizens in the California Club at the Waldorf Astoria.

Midway, Twain exclaimed: "This earthquake and fire transcends anything in human history, ancient or modern, but the same energy that built San Francisco in fifty years to be destroyed in a day, will build it again. Everybody is in a mood to contribute, from the hands of poverty up to those of the millionaire. But it is the poor man that gives most.

1906 earthquake. Photo courtesy of the San Francisco Chronicle

"The Salvation Army is the best means I know of to do this work. They are of the poor, and they know how to reach the poor. I have seen their work all over the world—always good."[7]

What Sister Sturdevant Saw

The Virtual Museum of the City of San Francisco has a prodigious website with a wonderful selection of letters and diary reports of the 1906 quake and fire. Add to them these first hand accounts from The Salvation Army's Sister Alice M. Sturdevant and others:

"From Mission Street to Bay, down Third, Fourth, Fifth, Sixth and Seventh to Nineteenth Street, there are nothing but squares of ashes not four feet high. The fire also burned clean up Clay, Washington, Sacramento, California and all those streets away out to the Presidio and all Chinatown and North Beach and Barbary Coast.

"Above the ashes, and dragging on the ground, are the wires; these point out where the streets have been.

"None were allowed in their houses and anyone starting a fire in a stove was shot by soldiers. Anyone rummaging anywhere was shot down. One man was caught disregarding the instructions and was shot and placarded as a warning to others.

"Corpses were left lying on Market Street after being shot.

"While we were camped on the hill, it was like one family—everyone divided up; even an orange was split as small as possible and the pieces given to the most needy and faint cases. Many were dazed and hardly comprehended it all. Anyone getting hysterical was called down quickly and shamed out of it.

"We don't know what it all means yet. I guess Shakes continue even now three and four a day."[8]

In the Middle of the Ashes, Sam Wood's Cornet

Provincial Officer Colonel George French, his wife and their overworked staff distinguished themselves again and again, said the Territorial Secretary's report following the first week. As far as he could see, the Army was first on the field with offers of practical relief. Oakland #1 Hall was immediately crowded with refugees. In one day, Adjutant Bradley and his helpers fed 1,600 hungry, homeless citizens there alone, in addition to those at Beulah Park. The Army uniform was respected and recognized everywhere. While others needed many military and other passes, the Army uniform was a pass in itself.

A stark picture stuck in their minds as officers visited ruins of the Provincial Headquarters. Fire had consumed the building. But there in the middle of the ashes stood the heavy metal safe . . . and part of Brigadier Wood's cornet.

And Where Were the Ensign's Shoes?

Salvation Army soldiers were camped together at the Presidio, where Ensign World and others were working. They had lost everything, but continued to help. The story is that Ensign World even lost her shoes. Not to worry. She was an adaptable sort. She found two mismatched shoes, simply put them on and laughed that under the circumstances she was quite sure one foot wouldn't recognize the other.

An Endless Procession

The Secretary's report included a series of sights from the street that were firmly etched in his mind. For five days he watched an endless procession of refugees trudging toward the ferry. Fashionable families pushed wheelbarrows full of their belongings. A woman alone pulled a trunk tied with ropes. Children carried their pets and favorite dolls. One old lady closely guarded two hens tied together by the legs.

Seeing the Army uniform, a fashionably dressed lady begged the officer to show her to a hospital and find water for the baby dying in her arms. It was a picture he couldn't forget.[9]

Across the street from the ruins of the Hall of Justice, a reporter for the *New York American* discovered the drum of The Salvation Army's Swedish corps. Apparently he knew the work of the Army. He reported that Salvationists were everywhere, working side by side with the Red Cross nurses.[10]

Union Square was swarming with refugees. As the flames spread, Skid Row men and women and immigrant families from the South of Market rushed into the fashionable part of town, disreputable, carrying what little they owned, looking for safety. But it was only temporary. By the next morning, the Square was a pile of ashes. Only the Dewey Memorial remained, and it's still there today.

These were the people The Salvation Army was working with even then—the homeless, people in trouble, people fleeing their pasts, people with nowhere to go but the streets.

Mechanics' Pavilion

Chaos reigned. In the streets, Salvation Army lassies, along with others, broke drug store windows and appropriated supplies for the wounded at Mechanics' Pavilion.

Lucy Fisher reported in the *American Journal of Nursing* that she had held a woman's crushed leg as they bandaged it. Ultimately, the leg was amputated. The surgeon thought the patient would recover, but with fires threatening, she was moved twice and later died from shock.[11]

This was Adjutant Anna Alleman Butler. The only Salvationist to lose her life during the fire, Butler was buried at Laurel Hills Cemetery where Kaiser Hospital is today and later moved near the Army's plot at Cypress Lawn.

Butler had lived at 913 Natoma, where heavy beams fell on her legs during the quake. Because of the approaching fire and chaos in the city, there's confusion about where she died. One report said she was moved to an emergency hospital where surgeons amputated the leg, then to Mechanics' Pavilion and next to Clara Barton Hospital.

The Coroner's Office record says she died at Dr. Simon's Sanitarium at 2344 Sutter. The cemetery interment record indicates she died at Mt. Zion Hospital and was buried 26 hours after the quake. *The Oakland Daily Herald* said she died at home.

"I visited the grave of our one earthquake victim, Mrs. Butler, who lost her life in saving that of her aged mother," said Commander Evangeline Booth when she arrived for a Memorial Day tribute. "To her was given the last casket in San Francisco and the last unoccupied grave."[12]

The combined San Francisco newspapers, the *Call-Chronicle-Examiner* of April 19, 1906, said there was terrific cooperation in the relief effort at the Mechanics' Pavilion. Matrons from Pacific Heights worked with neighbor ladies from down the street. Nurses, clergy, nuns and Salvationists worked side by side.

The Pavilion—a wooden building—served as a makeshift hospital from 5:30 a.m. until 1 p.m. That's when the infamous Hayes Valley fire sealed its fate.

Historians and earthquake cognoscenti, of course, know the fire by its popular name—the "ham and eggs fire." After all the excitement of the morning, a hungry San Franciscan decided to make breakfast, or maybe it was lunch, on a stove with an earthquake-damaged chimney. Enough said. A blaze broke out at 11:00 a.m. and spread uncontrollably. By the time it reached the roof of Mechanics' Pavilion, there was no hope for the building.

Volunteers from the street were recruited immediately and rushed in to move the 354 patients. The Salvation Army moved many on its small fleet of Industrial Home wagons, horse-driven forerunners of today's thrift store trucks.[13]

Bodies of the dead were cremated in the ruins.

San Francisco, the Movie

Picture it. San Francisco, April 18, 1906, as filmed on the back lot of a movie studio. A dance hall is filled with all-night party goers singing "San Francisco, Open Your Golden Gate" at the top of their lungs . . . then, without warning, at 5:12.05 a.m. the clock stops.

Earthquake!

In the 1936 Metro Goldwyn Mayer movie about San Francisco's big earthquake and fire, Clark Gable—bruised and bedraggled, but still debonair, of course—stumbles through the ruins of San Francisco searching for Jeanette McDonald, flame of his youth. Boyhood chum Spencer Tracy, now a priest, comforts a wounded patient. Next to him—compassionate but without a speaking part—a woman in black dress, cape and bonnet tends to the patient's needs. Surely it's a Salvation Army lassie. (Could it have been Elsie Alleman, the plucky Salvation Army officer who worked at Mechanics' Pavilion?)

A Lassie With Plenty of Pluck

In the early days of Salvation Army recruitment, application forms had always made a big point of looking for soldiers "with plenty of pluck." How do you measure pluck? And what was it anyway? Probably a little like today's version of "spunk" with some "stick-to-it-iveness" thrown in.

Although Horatio Alger's "how to succeed" novels were mostly written for boys, girls benefited from the advice, too. "Luck and Pluck" advised people to "go slow and sure" and "try and trust" as they worked to get ahead. Evangeline Booth commended at least one "plucky" little lassie following the quake.

She talked with a lieutenant who had been a heroine in the disaster (This was undoubtedly Elsie Alleman). The young woman had been at the Pavilion from the time it opened and vowed to see the last sufferer out of the doomed building. The stories she told were extraordinarily touching, said Booth. There was no water to wash the wounds. There were no anesthetics to deaden the pain.

Elsie Alleman continued working after her sister died. In one day alone at the General Hospital in the Presidio, she and Lieutenant Hynes served 1,500 gallons of milk in cupfuls, principally to mothers and children.[14]

Colonel George French said Salvationists' relief operations were met with praise and commendation. He felt that General Funston had been particularly considerate to the Army. Messages sent through an officer in uniform were given precedence over hundreds of others standing in line, including many wealthy people.[15]

As soon as refugees began to pour into Oakland, Salvationists were detailed to meet the ferries and local trains.

The combined *Call-Chronicle-Examiner* newspaper published on April 19 warned that everybody needed to be prepared to leave the city. It seemed inevitable that San Francisco would be totally destroyed. Still, thousands of residents stayed on in a dazed camaraderie that sometimes tried to make light of it all. An early-day precursor to Gertrude Stein predicted the gloomy prospect of leaving the city for less glamorous Oakland. The sign over one of the makeshift outdoor food stations warned:

"Eat, Drink and Be Merry, For Tomorrow
We May Have to Go to Oakland."

An early letter describes damage and the heat, even across the Bay.

Ensign Porter at the Army's Beulah Home Orphanage in Oakland estimated that damage to the Maternity and Rescue Homes must have amounted to at least $3,000 and the people in charge had no idea where they were going to get food. Even in Oakland, the heat there was terrible.

San Francisco had been burning for forty-eight hours.[16]

Staff Captain William I. Day reported in the May 19th *War Cry* that the Army had opened a department in the camp for lost and strayed children and also lost friends. State authorities, he said, had agreed to help.

In the confusion of sending refugee trains to the North, South and East, many children had been separated from their parents. Consequently in places like Oregon, Southern California, Nevada and Utah, there were babies and young children who couldn't tell where they came from or who their parents were, or even how they got lost. Some of the older ones knew the answers, but even then authorities could find the parents only through the slow, uncertain means of the long lists in the newspapers.

Parents come to The Salvation Army Camp at Beulah every day, inquiring for children they hadn't seen for a week.

Alma Chased Her Dog

In the same report, Captain Day pointed out a small girl clinging to her dog. She was Alma Jacobson, one of six children of Mr. and Mrs. Rudolph Jacobson, Germans who kept a fruit store at 1944 Mason Street. Fire had forced the family away from their home. They ended up at the beach. Then, in one of those instances of chaos in which everything happened all at once and no one could quite remember how, little Alma's dog ran away. She chased it. And when she got back, her relatives were gone. Captain Hynes took charge of the child and she stayed at the Army's camp while Salvationists checked newspaper lists of the lost to reunite the family.

Governor Pardee Instituted A Children's Reunion Camp at Beulah Park

Captain Day had glowing reports about the Beulah Park camp which was in the foothills, about three miles from Oakland City Hall. The gently sloping lawns and shade trees made it a beautiful spot. It was a particularly pleasant place, he said, for the children's reunion camp established by California Governor George Pardee at the suggestion of Governor George Chamberlain of Oregon. The way Day saw it, children of all colors, rich and poor, would have a good vacation in the country while waiting to be reunited with their relatives.

Newspaper reporters, sanitary inspectors, health officers and other visitors—including the William Randolph Hearsts—declared the camp to be the neatest and cleanest of the numerous camps around the bay. According to Day, Mr. Hearst said the Salvationists had the finest camp he had visited.

The camp served over 2,000 meals a day. Was there work to be done? Evidently, if someone asked for help, as many as forty people would respond. Volunteers swept the streets, scrubbed the dining room, peeled potatoes and did whatever else was needed.

Rebuilding

By the time of the earthquake, the Army in San Francisco had six mission halls for English-speaking people, one for Swedish and one for Chinese. There were two large shelters and also an Industrial Home at 273 Natoma where "out-of-works" could earn money for board and keep.

After the fire, it was all gone.

With the exception of The Salvation Army #4 Hall at 24th and Mission Street, and the house where the officers lived, every Army property in San Francisco burned to cinders.

The aggregate loss? Colonel George French later estimated it at nearly $150,000, "a large portion of it being assets of such a character as is impossible to cover by insurance."

His annual report for the year reflected the confusion and disarray of the first few days. "Officers and soldiers were scattered. Buildings were destroyed, records were partly burned, the rest unprocurable; funds were tied up in City banks."

On the whole the outlook was very gloomy.

But disasters tend to bring out the best in The Salvation Army. Within 24 hours, officers were serving meals to thousands of refugees. Working with the Oakland Relief Committee, they secured Beulah Park as a relief camp and for weeks, sheltered and fed from 200 to 800 people.

Oakland became the launching point for getting supplies into San Francisco. From all over the state, relief trains came with groceries, clothing and other materials. William Randolph Hearst's contribution arrived with Salvation officers from Southern California. Teams were secured and load after load was taken to San Francisco and distributed as quickly as possible. A branch Headquarters and Relief Station was opened at the only available place, the unburned hall at 24th and Mission Street.

Colonel French was proud of his troops, who lived for weeks in the relief camps, visited refugees, held meetings and tried to alleviate physical suffering. Captain Alleman and Lieutenant B. Hynes worked at Presidio General Hospital. For several weeks, he said, they were the only religious workers allowed to visit the injured treated there.

The Army—like the rest of the city—started stacking bricks, removing ashes and rebuilding rapidly. By the middle of July, the headquarters building on Mission Street was completed. It was the hub of administrative activity. It was also a shelter. The Army's role in helping to rebuild the city provided jobs, beds and meals for homeless men.

Looking for property in the new business district took time. Rents were excessive and space was limited as organizations were forced to relocate. Finally, in late August the Army leased the upstairs at 1702½ Fillmore Street, on the corner of Post. This was the "New Congress Hall" designed to take the place of Corps # 1, 2, 3, 6 and 9 which had been burned.[17]

Californians Have Spirits Like Corks!

Perhaps the words of Commander Evangeline Booth in an early issue of *The War Cry* tell it best. Part description, part pep talk, her message was upbeat and positive.

"The natural hopefulness, I was going to say sunnyness of the Californian disposition, has stood them in good stead; all their troubles and privations have not been able to keep them under. They have spirits like corks and come up with a bounce every time."[18]

Resiliency was the word. Within the year, the city had thousands of new buildings. Nine years later, San Francisco's Panama-Pacific International Exposition of 1915 featured exhibits from 29 states and 25 foreign countries.

Nearly 20 million people came.

What Made Those Doughnuts Taste So Gosh Darn Good?

Mom and apple pie, a postcard from home, and eventually the taste of a homemade doughnut. That's what soldiers away at war missed the most.

John Milsaps knew it. When he went to the Philippines during the Spanish American War in 1898, he was the first Salvation Army military "chaplain" recognized by the U.S. government. He had little to offer—just a good heart and a few supplies. But he knew boys away from their families for the first time needed a place to relax and write letters to folks back home.

Mother Emma Goldthwaite knew it. In 1904 when she saw off U.S. military transports leaving the Golden Gate, she wasn't simply standing on the dock waving goodbye or welcoming the troops home. Her ministry was practical . . . and on at least one occasion a little perilous.

In 1919 a transport steamer returned from Siberia. She was 60 years old, not exactly a youngster. But she was determined to get on that ship.

The crew was flabbergasted. As they watched, she climbed a rope ladder hand over hand to board the ship mid-stream. Once on board, she distributed telegram blanks and collected over 100 messages to dispatch to worried relatives back home.

The War to End All Wars

Evangeline Booth, The Salvation Army's National Commander, also knew the boys "over there" in World War I needed mothering. Her first thought was bandages. That was easy. How could you ever have enough bandages? During her "Old Linen Campaign" in November 1914, workers stripped and rolled lengths of sterilized linen into bales and bales of clean supplies.

Under her supervision in 1917, the Army opened war huts or "rest rooms" at military bases throughout the United States.

Finally, she hit on yet another idea. And this was one for the history books. She advised General John Pershing that her peaceable Army would be a tremendous morale booster on the front lines. They would "mother" the troops, she said. They would endure any hardship and report for duty immediately.

"But I already have an Army," he gently reminded her.

"Not MY Army!" she countered. She was determined to put her troops into action.

The fact is Pershing was especially inclined to listen to the Army. When he was a young colonel temporarily stationed in San Francisco, his wife and son had died in a fire. The Army sent condolences when very few people in town knew him. It was a small act of kindness. But he remembered.

So he finally broke down and authorized the first group of Evangeline's special troops to go. In fact, he even designed their uniforms. Eleven people were commissioned—one couple, four single women and five single men.

Doughnut Girls

One girl made the first doughnut. One girl fried it. Another had the idea in the first place. Whichever young Salvationist was the very first lassie to fry a doughnut in the World War I trenches sometimes seems to depend on which newspaper article you read in which part of the country.

The remarkable thing about the women who stand out in the Army's doughnut saga is that many continued in community service for the rest of their lives. And most lived very long lives.[19]

Geneva Ladd Staley was in the very first contingent. In Stanislaus County, California, she celebrated her 99th birthday after serving 64 years as a registered nurse. The secret of her longevity? It was just one sentence and straight to the point. "I'm a tough old bird."

Remembering the war, Staley spoke of one officer who warned, "These girls are here to help us. If I hear of anyone behaving in a way which is any other than that of an officer and a gentleman, he will die!"

Alice McAllister Baugh, who later lived in Marin County and Santa Cruz, traveled with her sister Violet, guitars in hand, to entertain the troops. When they found how starved the soldiers were for home cooking, they first tried making crullers in a galvanized can. Later they made do with the simplest of supplies—a grape juice bottle for rolling pin, tin cans to cut the shape, a coffee percolator tube to make the hole. The day a line of 800 from the 26th Division lined up for the first 150, they knew they had found their calling.

As for the actual first doughnut prepared in the mud, cold and rain of the frontline trenches, Margaret Sheldon provided the recipe for posterity in her endorsement of KC Baking Powder. According to reports, in 1917 on a jerry-rigged stove using baking powder, flour, sugar, canned milk, water, lemon, vanilla and mace, she and Helen Purviance (who was promoted to Glory as a lieutenant colonel at age 95) made between 2,500 and "9,000 doughnuts a day."[20] The number of doughnuts and the recipe changes, depending on who tells the story.

"My Doughnut Girl"
The Song. The Girl in Charge.

Stella Young was the lieutenant at the Army's first doughnut canteen in France and admitted that frying doughnuts was not her idea. But she was in charge. It was her decision to continue. She also became the poster girl whose picture carrying a tray full of doughnuts was widely distributed. A song, "My Doughnut Girl," was written to honor the work.

After she returned home, Captain Young took to the lecture circuit. Vividly, she described a village shelled so badly that they had to camp fourteen days in the woods. Did that stop the doughnut brigade? Absolutely not. They stood over the stove for hours at a time. And when they did look up from cooking, they said, "It seemed as though the whole American army came by . . ."[21]

Young went back to Europe in World War II to direct a serviceman's club. After 40 years of service she finally retired, but her early training stuck with her. Wherever she lived, she gave utility kits to every local person leaving for military service. She was promoted to Glory at age 92.[22]

Doughnut Girls wrote graphic details about their service during the harsh winter. One said when the boys came in from the trenches they could literally stand their coats on the floor. They were that cold, frozen stiff. It was an honor to make coffee for those boys, remembered the girls. In a way, it was as though they were doing it for their own brothers, doing for them what their mothers would have done had they been at home[23]

Found in the Attic

When Major Trish Froderberg received a cache of memorabilia from relatives of one of the first Salvation Army officers to go to France in World War I, she carefully separated everything into labeled manila envelopes, gingerly unfolded old newspaper clippings and methodically laid out the photographs. The history was intriguing. But even more than that, the depth of commitment moved her deeply. The story of Salvation Army officers William and Beatrice Hammond began as husband and wife left New York in November, 1917, waving goodbye to their 11 year old son on the dock. Salvationist families would take care of him while they were gone.

An undated newspaper article described the young adjutant's service. "While the 87th Division occupied the town, she worked for seven weeks twenty hours a day frying and serving doughnuts and managing The Salvation Army hut. She served the doughnuts in a barn half a mile from the hut and close to the German lines. Only at night could she reach the barn. During the St. Mihiel drive, according to the article, Mrs. Hammond persisted in remaining at her post through a 72-hour barrage, despite orders for her to leave. She was of tremendous assistance when the wounded were brought back. In one week, she had just seven hours of sleep. At the end of the fighting, Mrs. Hammond was at Cheppy, in the Argonne, reaching there after a 17-hour drive in a motor truck and a 22-hour fast."

We can get a glimpse of her feelings for the work from faded notes for a speech she gave on her return, perhaps in San Francisco where she served for many years: "I am so glad that I belong to organization that gives a woman an opportunity for service as well as men, and if I cannot do great things, I will try and be faithful with the little things that I can do, so that it can be said of me when my work is finished 'She hath done what she could.'"

Recipe for Doughnuts
Printed in K C Baking Powder Flyer
With Margaret Sheldon's endorsement of the product

Flour, 2 cups (sifted)
Salt, ¼ level teaspoonful
Sugar, ½ cup
Mace, ¼ level teaspoonful
K C Baking Powder, 1 level teaspoonful
Shortening, 1 tablespoonful
Milk, ½ cup

Reserve one-fourth cup of flour for the board. Sift the remaining one and three-fourth cups of flour with the baking powder three times, and set aside. Cream the shortening, sugar, salt and mace, add the milk and stir, then add the flour–baking powder mixture. Work into a soft dough and roll on the floured board into a sheet ¼ inch thick. Cut in the desired shape and fry in deep fat, turning the doughnuts frequently. The fat should be sufficiently hot to give the doughnuts a rich golden-russet color in three minutes. While hot, roll in sugar. This recipe will make about fifteen good size doughnuts.

Another Version

4 cups flour
¼ tsp grated nutmeg
1 cup sugar
1½ tsp salt
¼ tsp cinnamon
1 cup milk
½ tsp butter
4 tsp baking powder
1 egg

Put flour in shallow pan; add salt, baking powder and sugar. Rub in butter with fingertips. Add the well-beaten egg and milk and stir thoroughly. Toss on floured board, roll to one-inch in thickness, shape, fry and drain.

And Now They're Available at the Corner Store

In 2003, The Salvation Army in Seattle, Washington, lent its name and recipe to "The Salvation Army Famous Doughnut," a commercial offering baked in Clackamas, Oregon, and sold through Fred Meyer stores in the Pacific Northwest. According to the label, part of the proceeds go to the Army in the "battle being fought against the unfair penalties of poverty: hopelessness, homelessness and hunger."

Forget the Brass!
Those Doughnuts Were for Enlisted Men Only

Word soon got around that Salvation Army lassies were in France to help enlisted soldiers, the men in the trenches, not the company brass. They had no special privileges. They worked with the men under the same conditions, ate the same food and went without pay. When invitations were offered to join the officers' mess, the women politely refused and took their places in the chow line. Seeing this, soldiers stepped aside to let them in.

"They Don't Put On No Airs"

Letters of commendation praising the Army's unique service arrived regularly at headquarters. They came from heads of state, generals, governors, colonels, chaplains and, most of all, the enlisted men.

"You see, Judge, the good old Salvation Army is the real thing," said one. They don't put on no airs. There ain't no flub-dub about them and you don't see their mugs in the fancy magazines much. Why, you never would see one of them in Paris around the hotels. You'd never know they existed, Judge, unless you came right up here to the front lines as near as the Colonel will let you."[24]

A Letter from Somewhere in France

"I do not know what we boys would do if it was not for The Salvation Army," a soldier wrote to his mother. When he got home, he declared, he would exchange the United States uniform for The Salvation Army uniform and added, "I know, Ma, that you will not object."[25]

How Many Doughnut Girls Were on The Front?

However many Doughnut Girls there were in France, the numbers were certainly not so large as the mythology that has built up around them. The late Colonel Henry Koerner noted the comparison of numbers from several Salvation Army historians.

It's generally understood that as few as 250 Doughnut Girls from the United States were sent to France. Herbert Wisbey, Jr. added that another 400 men and 150 women workers were secured on the other side. According to Grace Livingston Hill, writing in 1919, "1,507 Salvation Army officers devote their entire time to religious and social work among soldiers and sailors." She listed 45 chaplains serving under Government appointment.

Salvation Army historian Edward McKinley compares that number to over 10,000 YMCA workers. At first the Army's work was very similar to that of the YMCA, which was officially responsible for welfare work. The Red Cross's official assignment was relief work.[26]

Once those doughnuts started frying, the Army was in a league of its own.

Frying doughnuts, holding pie contests, making pancakes, frying eggs for breakfast, baking birthday cakes and letting soldiers slip in to scrape the bowl clean just like they did back home—these were the sights and smells of the kitchen.

But there was more to Salvation Army service than the doughnut. First there were basic household chores like darning socks, sewing on buttons, mending coats. In field hospitals, the lassies washed wounds and wrote letters to soldiers' sweethearts back home. At the men's requests, they forwarded paychecks to their families through Salvation Army units in America. Going into battle, soldiers left letters and thousands of dollars in cash or valuables with the Sallies to forward to their families just in case they didn't make it back. They say that if a soldier needed "jawbone," or credit, it was given without hesitation.

Pay it forward, said the Sally, making a loan. Pay it back when you can. Give it to the next Salvation Army unit you see.

And How Much Did That Doughnut Cost?

Although there was a small charge to assist with costs for doughnuts in some places, it seemed casual and, in many cases, simply not a factor. The same holds true today. Whenever you see a doughnut being offered at a Salvation Army canteen, there's no charge. Army officials are strict about that. A kettle may be placed somewhere nearby if you feel moved to make a donation, but in the canteen money does not pass hands.

Actually, free doughnuts make far more sense. A small contingent of Army lassies gave away doughnuts during World War I. A positive feeling of good will has followed ever since. The Red Cross sold a few donated cigarettes during World War II and has never quite lived it down.[27]

Male Salvationists served in France as well. In fact there were more men than women. The women just got better press.

During his lifetime William Booth had adamantly said the Army was a movement, not a denomination. But in September 1917, the Judge Advocate General of the War Department declared that functions of Army ministers seemed similar to those of the clergy of any other church. His judgment made it officially possible for Salvationists to serve as chaplains.[28]

The Home Service Appeal Eventually Led to Service Extension

The war was over. Never again would The Salvation Army be considered simply a ragtag group of musicians standing on the street corner singing hymns. Nationally, a major campaign was launched to support its work.

The enthusiasm it generated became the foundation for Service Extension work now extended to communities throughout the country to locations where there is no Salvation Army corps community center.

Back in 1919, a well-organized, widely-based drive was run from the headquarters at 633 Market Street in San Francisco. Telegrams to local donors from the chair of the executive committee John McNab in March, 1919, described "noble men and women exhausting themselves in physical effort saving the suffering and downfallen from despair."

"It wasn't doughnuts all the time," said an original poster that now hangs in the development department of The Salvation Army's divisional headquarters at 832 Folsom Street in San Francisco. "Sometimes it was bandages and first aid kits under shell and machine gun fire, binding a wound, taking a last message, trickling water down a parched throat . . ."

"Ask the next soldier boy you see with the overseas cap and the Sam Browne belt—ask our boys of the 67th—what about The Salvation Army . . ."

"They didn't pick out the soft snaps, away back behind the lines" is the message of another poster from that campaign. "They were up in front, whether they had doughnuts or baked us a batch of home-made bread. Say, how they did it, I don't know. It made us think of ma and sis and shucks, the girl back home waiting until we got back—if we ever did."

A New War and The USO

During the summer of 1940, The Salvation Army invited Catholic Community Services, the Jewish Welfare Board, the YMCA, the YWCA, and the Travelers' Aid Society to join a united effort to offer services to the country's armed forces. On January 31, 1941, leaders of the six agencies formed the United Welfare Service Organization (USO). Commissioner Edward Justus Parker, who was National Commander at the time, made it clear this was not a new organization. It was a uniting of six existing organizations that had years of experience in the social service field.[29]

By March, the Army was operating 46 camps, 27 of them in the South and West. In addition to clubs it operated for the USO, Salvationists also ran emergency service centers called "Red Shield Clubs," with showers and sleeping facilities.[30]

The Salvation Army USO on Market Street.
Postcards on these two pages courtesy of www.sacollectables.com.

After Pearl Harbor

Henry and Marie Koerner were stationed in Hawaii on December 7, 1941. Henry wrote about taking his Salvation Army canteen on its first nighttime run into the darkness of an island stunned by the day's events. It was so quiet, so still, so "inky black" that it seemed like midnight, he remembered. In reality it was only 7:30 p.m. The streets were vacated. Without a police permit, people were ordered to stay inside.

The Army, however, had a specific duty—to take food to the troops. As they came to their first stop, they heard a voice from out of the darkness. "Halt! Who goes there?"

In the dark, it was hard to tell who was speaking. Koerner remembered the caution with which they approached these first soldiers, who were armed with machine guns.

"It's The Salvation Army USO with coffee and doughnuts," he replied.

"Advance and be recognized" came a voice from out of the darkness. In the beginning, the soldiers answered back with textbook military protocol.

Later, once they had become old friends and once the doughnuts came to be expected, the welcome was much friendlier.

"Advance and be devoured!"[31]

The Koerners returned to the United States when the curfew, blackout and military government were still in effect. Without an escort, their troop and cargo ship zigzagged home across the Pacific in 1943.[32]

In San Francisco Salvationists did canteen duty many nights during blackouts. Even as window shades were down and the city was pitch black, Salvation Army volunteers silently drove up the winding road to Twin Peaks to take food to troops stationed on alert.

On May 29, 1942, the Army's national commander Commissioner Edward Justus Parker dedicated the USO's 700th club on "the great main thoroughfare of the Golden Gate City, San Francisco." According to his memoirs, thirty-seven radio stations broadcast the dedication ceremony.[33]

The Salvation Army's USO in San Francisco was a second-floor walkup at 989 Market Street at Sixth, across from the Fox Warfield Theater—a refuge for young boys fresh off the farm who had come to sign up for war duty.

San Francisco was swarming with servicemen and the sweethearts they would have to leave behind. Young women from all over the country came looking for their beaux whose addresses they knew only as APO San Francisco.

Since the city was the last port of call before men were shipped overseas, there was a sense of urgency and always, always, there were letters to write, packages to mail and even weddings to perform.

As ordained ministers, Salvation Army officers provided marriage ceremonies as well as counseling. Like other San Francisco USO agencies, this was a home away from home for the troops. Ask those who served what they remember most. Colonels Henry and Marie Koerner remembered homemade pies, finding rooms for families, and, most of all, wrapping and mailing packages and personal voice recordings to send to the folks back home.

Henry was responsible for overseeing all Salvation Army USO clubs and mobile canteens in the western states from 1943 to 1947.

It was usually midnight before Salvation Army officers left the building, according to Arlee Lansing, whose parents were in charge. Loaded with sacks of mail, they stopped at Rincon Annex every evening. Volunteers from throughout the city helped.

One night, Dorothy Koerner Bruggman was asked if she had heard any news about her husband who had been listed as missing in action in France.

"Yes," the 26-year-old replied quietly. Just that day, he had been confirmed dead. "But how could you possibly come in to work now?" her friend asked.

The answer was a sign of the times. "What else could I do? I couldn't just stay home and do nothing."[34]

A Terrible Mix-up

Just as Mother Emma Goldthwaite dispatched telegrams to families at the turn of the nineteenth century, Captains Victor and Ardys Newbould performed the same service during World War II, meeting ships as they docked and sending word ahead to parents that their sons and daughters had arrived back in the United States safely.

One of their letters arrived on the east coast *after* a family had already received official word their son had been killed, *after* they had already held services at the synagogue for him.

The family, still in mourning, was dismayed by the Salvation Army letter claiming their son was still alive. Obviously this information was incorrect. By simply checking with the government they could have avoided this dreadful mistake.

Newbould was momentarily devastated. Immediately he reached for the phone, found out the government was dead wrong and called the parents back. Their son wasn't dead. He was alive in a San Diego Naval Hospital mental ward. The family was stunned. Newbould's wonderful information made them Salvation Army believers.[35]

Long Johns and Washing Machines

During World War II, Captain "Sammy" Bearchell was ombudsman to U.S. Navy ships. She brought the sailors long johns and homemade cookies, wrote their families and arranged for washing machines on board the ships. For this she was named to "the Royal Order of the Alligators" honoring people who helped the landing craft infantry of the Navy. Did she make an impression? One of her sailors looked her up 40 years later! After retirement, she served as San Francisco Family Services Director for many years. Like so many of her Salvation Army sisters, she lived to a ripe old age—just five months shy of her 99th birthday.[36]

**During the Korean Conflict
A 36-inch TV Screen**

Twelve hours a day, seven days a week during the years of the Korean conflict, The Salvation Army Canteen Kitchen (SACK) in the San Francisco Ferry Building was open to military personnel en route to duty. Coffee and doughnuts were standard fare.

But there was much more: entertainment passes, hotel and travel information, sightseeing, transportation, missing persons search, letter writing, home hospitality, home-cooked meals with San Francisco families and what was evidently a big selling point at the time . . . a 36-inch TV screen!

Five scrapbooks filled with memories help to put a human face on the story of the SACK Canteen in San Francisco during those years. Helen Jackson, public relations director, filled the pages with keepsakes from hundreds of military personnel she met—thank you letters, Christmas cards, postcards, wedding invitations, birth announcements, baby pictures, family snapshots, correspondence to and from people on overseas duty . . . and sometimes letters from the same soldier five years later. A few samples from her scrapbooks show a cross section of people who used the Army's services and a snapshot of earlier times.

Just Like Grandmother

"Dear Mrs. Jackson: My son said you reminded him of his Grandmother and he thought there was no one like her so he holds you pretty high."

Just Like Mom

"Dear Mrs. Barry: I want to thank you for a most wonderful evening at your home Christmas. My buddy and I were rather in the dumps until we hit your hacienda—then we remembered what a home was like. You and the others were most kind to have me out—I realize it was a lot of trouble—I remember what Mother goes through just for a couple of tables of bridge."
—*Joe's letter to a volunteer host*

Wonderful Creatures

"Dear Mrs. Barry, Linda, and Bob: It is perfectly incomprehensible how you have prepared such an elaborate feast under the circumstances but I have long ago ceased to wonder at the powers of a woman and, of course, when you put two of these wonderful creatures together, perhaps it is not so difficult to comprehend after all."
—*Sincerely, I am your friend*

Four Months Without A Letter

"Dear Sir: I am writing in about my boy in the navy. I haven't heard from him for 4 months and I cant (sic) understand why I dont (sic) hear from him. I sure would be gratified to yu (sic) if yu (sic) could find out what is the matter that he dont (sic) write, and would let me know."

Thanking yu (sic) kindly
—*Letter to Major Harold Barry, June 29, 1953*

Thanks for Being So Nice to My Daddy

"Mommy, tell that nice lady thanks an awful lot for being so nice to my daddy, and send her a big hug and kiss from me."

*Forever grateful from
Wenatchee, Wn*

On the Way to Vietnam
In the Middle of the Night

From the very beginning, Salvationists seem always to have known how important it was to be at the dock when the ship sailed, at the train when the whistle blew, at the bus no matter what time of day or night it was ready to roll.

"My first contact with The Salvation Army was at 3:00 in the morning," remembers Major Ron Toy about reporting for duty at the Oakland military induction center. The only one seeing him off for service in Vietnam, other than his father, was The Salvation Army.

"Salvationists, I don't know if they were soldiers or officers, were standing by the buses, handing out kits and wishing us well. Imagine, three o'clock in the morning! And a Salvationist was there at that time to see us off!"[37]

A Cup of Cold Water

Ashes cover the landscape. Every tree is a charred stick. It's a grim sight and The Salvation Army often sees it.

Families returning to the ruins of their expensive Oakland Hills homes in 1991 gasped and cried. Occasionally, pieces of their past surfaced. But, for the most part, everything was gone.

Even before homeowners were allowed back into the area, Salvation Army mobile field kitchens had been on duty. As the hills went up in flames, one inferno after another, the air was so thick with smoke firefighters couldn't see any farther than a few yards.

Then, as they pulled back at the change of shift to catch their breath, relief workers took off their helmets and stopped to receive a gift the Army considers one of its most precious services at the very moment it's needed most.

Food, yes, nourishment, yes, a listening ear, yes, and sometimes simply a long, cold, extremely satisfying first gulp of water to quench the thirst.

Think of the imagery. A cup of cold water.[38]

It seems so simple.

Maybe that's what Salvationists love best about it. It represents the smallest acts of kindness, the kind money simply can't buy.

Easy to give. Incredibly satisfying to receive.

Pragmatically, Army mobile field kitchens provide breakfast, lunch, dinner and snacks at all hours to firefighters during their most grueling work. During clean-up, they provide cleaning supplies, buckets, shovels, gloves, boots and bleach. During the immediate crisis, and long after, they minister to people whose lives have been uprooted.

Symbolically, that cup of cold water represents a ministry of presence, being there to help in any way possible. It's what the Army does best.

The Day Juan Built His Own House

The Guatemalan earthquake of February 4, 1976, was a big one—7.5 on the Richter Scale—leaving 23,500 dead, thousands injured and a million homeless. The Army jumped in with both feet to help, giving tons of supplies, medicines and food, missing persons assistance and ultimately a cement block-making machine that would build an entire town.

Getting the machine from Vancouver, Washington, to Guatemala proved to be quite a drama. The machine was too heavy to go by plane. Officials decided to put it on the next ship headed south. But they missed the boat.

Major George Duplain rushed from San Francisco to

help. He put the machine on a truck headed for the Bay Area and raced down the coast to meet the ship before it left on the next leg of the trip.

Once in Guatemala, the block-making machine, which could turn out 2,000 cement blocks an hour, became the foundation of an industry that literally rebuilt the town of Tecpan from the ground up.

People learned a new trade and, in the process, built 524 new houses.[39]

"We're Going to Mexico!" Said the Colonel As He Watched the Evening News

"Alert the media.

"We're going to Mexico. And we'll need some money!"

The orders came from Lt. Colonel Ray Robinson, Divisional Commander of Northern California and Nevada. The tone of his voice was resolute. He was ready to move.

An 8.1 earthquake had hit Mexico City at 7:20 a.m. on September 19, 1985, destroying Benito Juarez Hospital and killing more than 600 immediately. There were rumors of a *tsunami* (tidal wave) to follow.

Robinson made up his mind quickly. The Senior Meals Program had already given a picnic for 2000 seniors that day at Sigmund Stern Grove. At 7:00 p.m. his adrenaline was churning. He was going to Mexico!

By the 10:00 News, word was out. At 7:30 the next morning, KGO-Radio and Bay City News sat in on a planning session that would continue for the next two days.

"We'll need a plane," said Robinson.

"He needs a plane," said a reporter to his listening audience.

And within an hour, several small plane owners had responded. So had more than 100 volunteers and donors offering medical supplies.

Robinson sighed.

"We'll need a bigger plane."

Who would go? The first priority, of course, was multi-skilled Spanish-speaking emergency room personnel. In the end, general practitioners, pharmacists, nurses, a construction team and translators headed the list. All day Thursday and Friday into the night, the team came together. It was an astounding effort. On Saturday morning at noon 49 volunteers and eight tons of supplies plus reporters from KRON-TV, the *San Francisco Examiner* and *Hayward Review* flew out of SFO on Butler Airways. Rental of the chartered plane was $35,000. By Monday morning, San Francisco donors had already covered the cost.

It was the first and largest relief team to arrive in Mexico City.

Finding little they could do at the hospital site, (it was totally destroyed with the bodies inside) the Bay Area medical team moved to Colonia Morales, one of the poorest barrios in the city. It was Sunday morning and the local Salvation Army was in church when the San Francisco team arrived.

"Well, there's a time for prayer. And a time for work," chided the gringo colonel. "I have a team of 49 people here. Let's go to work."

And there, in a local Salvation Army building serving a neighborhood with few other services, the North Americans set up a makeshift clinic for people whose already unmet health needs were exacerbated by the quake.

Volunteers slept on the floor, carried water from outside, jerry-rigged examination tables out of boards placed across pews and sent medical teams out into the community. They treated patients with diarrhea, sore throats, cuts and congestive heart failure, accompanied them in ambulances to local hospitals, delivered at least one baby and treated infectious wounds. They answered calls to perform surgery in local hospitals, repaired broken medical equipment, inspected earthquake-damaged buildings and helped in the continuing search for survivors.

In the months that followed, eight relief teams from The Salvation Army's Western Territory served in Colonia Morales. KRON-TV virtually adopted the project and contributed all its footage to a Salvation Army film about the effort. More than $900,000, the largest portion of which came from San Francisco, was raised to build a large permanent clinic that still serves that community.

Christmas in the Barrio

The camaraderie of the Mexico medical volunteers went far beyond the initial effort. They were a team now. At Christmas, a small group returned with $25,000 in donations to purchase toys and provide a 'Christmas in the Barrio" party for 2,300 children in Colonia Morales.

On Christmas Eve—in a vacant lot under a bridge

on cement flooring made from a block-making machine donated by KGO-TV—the team distributed gifts and food to children still living in tents. Balloons, cotton candy and kids were everywhere. Even the local police pitched in to help.

At the end of the day, not a single toy, ornament or scrap of tinsel was left. A woman came late, carrying a small child in her arms. There was nothing to give her, only the smallest piece of glittery decoration attached to the chain wire fence. She took it and smiled as if we had given her gold.

Later, in that hulk of a warehouse structure, local Salvationists shared their dinner, laughter, hugs and thanks with the visiting team. Later still, volunteers walked through earthquake-damaged streets to midnight mass.

It was a magical day.

A Place Sadly Called Paradise

A torrential thunderstorm hit El Salvador in October 1986 following a 5.4 earthquake that killed nearly a thousand people. Those in the poorest barrios were hit hardest.

"It's raining in El Salvador," we informed the Bay Area press. "People are homeless. They're sleeping on the streets. And it's raining."

This, of course, was not an entirely new phenomenon. In a country divided by class, torn by war and anguished by poverty, there were always people sleeping on the streets and during the rainy season. The disaster relief team that had gone from San Francisco to Mexico City the year before was eager to help in San Salvador as well. From the Ilapongo airport they rode past the ditch where four Maryknoll nuns

Joe Posillico, (far left) who became Divisional Commander in 2004, served ten years as Divisional Emergency Disaster Coordinator. In 1985 and again in 1986, he brought together a remarkable team of volunteers who helped in Mexico City and San Salvador following devastating earthquakes in those cities. El Salvador Consul General Gloria Ayala de Gavidia is shown here with some members of the team. Photo by Kazuhiro Tsuruta.

had recently been murdered and headed past the cathedral where Archbishop Oscar Romero was assassinated. The group pulled up in Colonia Paraiso, a desperately poor barrio ironically named for paradise. At that point, civil war had rocked the country for seven years.

Just hours before, a bulldozer had ripped through the area, reducing fifty earthquake-shattered houses to rubble. In their place, The Salvation Army built a tent city, a huge MASH tent and field kitchen surrounded by a dozen smaller ones in which to cook and distribute medical assistance.

Hundreds of houses ringed the periphery. It's as though the bulldozer had simply gutted out a huge circle from a jumbled mass of structures piled one on top of the other. From somewhere in the distance, a radio was playing.

By nightfall, it was pouring!

Torrents of water filled the roof of the tent so dramatically that it sagged just inches above volunteers sleeping on cots beneath. Surely just one more drop would have sent the water crashing down on their heads.

Salvation Army volunteers had set up a dirt-floor clinic as soon as they arrived. Nurses, pharmacists, construction workers, electricians—people with a good sense of what was needed, both medically and personally—they provided basic first aid and medicines to long lines of people who had been assembled since early morning.

For the most part, the volunteers were Hispanic. Their language was Spanish, so the community welcomed them. Neighbors cooked huge pots of beans and rice, offering assistance and friendship wherever possible. Children emptied garbage cans and ran errands. They were very proud to be part of the effort.

The next morning after the rain, a group of them ran wildly into the camp to report a dead baby left on a trash heap during the night. It was a 6 to 7 month fetus. Immediately, a surrogate family emerged. A neighborhood woman washed and dressed the body. A volunteer prepared a coffin. Another donated the $150 *colones* ($30) for burial.

Ana Maria Robles organized a basic pharmacy in the rubble of San Salvador. Photo by Kazuhiro Tsuruta

Throughout the day, patients waited patiently. A ten-year-old boy described what seemed to be a simple fever and cold. "When did it begin?" asked the doctor. "The day my mother died in the earthquake," he said. A small child had been scalded by a pot of beans during the tremor. A seventy-eight-year-old woman had taken a five-hour bus ride from the Guatemalan border because she didn't have twenty *colones* ($4) for a prescription.

A 104-year-old woman with a broken hip smiled coquettishly as her friends made sure the reporter knew of her advanced age and moved her to the front of the line. She, like hundreds of others, had come many miles for medication. *San Francisco Chronicle* photographer Steve Ringman photographed that feisty old lady.

In the first two weeks, The Salvation Army team treated over 4,000 patients and distributed 400 tents, 2,200 blankets and over 25,000 pounds of food. At Christmas, along with "Caring for Children," Emergency Services Director Joe Posillico and the Army's number one translator and later public relations assistant, John Garcia, took 800 Bay Area teddy bears to a local orphanage

The rains continued. And eventually, the political climate made it difficult for the clinic to survive. But for a very short time, gently and with great care, The Salvation Army tried to heal a few of the wounds in a place sadly called Paradise.

Volunteers had gone because they had a deep tie to their heritage. With great tenderness, they dispensed basic medicine and reached out to people with headaches and high fevers and broken toes.

And when they came home, they were never again quite the same.

Luisa Almeida-Oetting (right) brought nursing skills to a makeshift Army clinic. Photo by Kazuhiro Tsuruta

Other Disasters/Other Times

The late 1980s and early 1990s were heady times for disaster services. As word got out that The Salvation Army was willing to travel to disaster sites, calls began to come regularly for assistance in other places.

Dianne Feinstein, then mayor, mentioned Chernobyl. "Oh no," said Colonel Robinson to Joe Posillico. "We can't stretch our resources that thin." Later, intrigued by the possibility, he backed off a little. "What did she have in mind?"

A news anchor was quizzical. "I don't understand. Are you guys now doing what the Red Cross does?"

And for several years we were. A fleet of Salvation Army canteens across the state responded regularly to California forest fires and floods. In the middle of the night, in the middle of the forest, we drove through fire lines to hand out sandwiches, fruit and coffee to relief workers in remote locations. Often there was no one to relieve them. There was no way for them to get in to Red Cross feeding stations back at camp.

"Hi, where are you from? When was the last time you ate?" (The answer was often several hours before.) "Have a hamburger. Take some fruit. Thank you for coming to California to help us fight this fire."

Mexico City, El Salvador, Loma Prieta and Northridge earthquakes, Hurricanes Georges, Mitch, Andrew and Hugo, the Oakland Hills Fire, Mississippi floods, Iniki, Acapulco, Bangladesh, Kosovo, Gujarat—disasters seemed to come one after another during those years. And they often came in September and October. During this time of intense community response to other people's hurricanes, other people's fires, The Salvation Army often sent volunteers and collected funds for assistance in devastated area.

The Salvation Army serves in many countries. By 2004, the number had grown to 109. Ideally, local and neighboring units are able to take care of disaster needs. Unless there's a window of opportunity open for free transportation from an airline, it's generally too expensive to send goods.

Money, of course, is always needed. Any agency will tell you that. Cash donations make it possible to purchase in bulk at wholesale prices, help the local economy and purchase selectively those items that are most needed. Overseas donations are forwarded through The Salvation Army World Service Office (SAWSO).

Most important, if you designate where money is to go, that's where it goes. If there's no Salvation Army personnel in a country, the organization forwards earmarked donations to agencies it trusts. When you send a general, undesignated donation to The Salvation Army, you're trusting its best judgment about how to spend the money.

Loma Prieta—Disaster in Our Own Back Yard

At Divisional Headquarters in San Francisco, tucked underneath a desk during a 7.1 Loma Prieta earthquake in the Bay Area in 1989, you could hear cabinet drawers opening and then slamming shut again. The Xerox machine literally moved three feet away from the wall.

All ten floors at the Silvercrest Residence for Seniors were without lights. Volunteers climbed the stairs, going from apartment to apartment to check on the residents.

A blind woman sat in a rocking chair, worried that she had spilled a bottle of cranberry juice. The floor was sticky and strewn with broken glass. It was pitch dark. She was used to that. The volunteer wasn't.

The woman tried to be helpful, holding a flashlight and pointing it in the direction where the volunteer was trying to pick up the glass. But she was rocking back and forth, back and forth. In rhythm with the chair, the light also went up and down, up and down. In the dark, trying valiantly to follow the bouncing light—the volunteer bounced up and down, up and down, to clean the mess.

Recovering alcoholics at Harbor Light made sandwiches in the middle of the night. Breakfast was arranged for relief workers. Salvation Army officers helped at the Oakland morgue. The Adult Rehabilitation Center immediately sent a truckload of clothes to the Marina for people unable to get back into their homes for a change of clothing.

The Salvation Army had moved into its new Folsom Street building just a few months before. Going from floor to floor to assess damage on the night of the quake, someone set off an alarm. The building was new. Who knew how to turn it off?

We worked all night. The alarm continued relentlessly. San Francisco firefighters were busy all over the city. At dawn a reporter from WMAQ Radio in Chicago called to get a report on the Army's response. With a siren wailing in the background, Major Gordon Helms described the problem to the Chicago audience. Fortunately a firefighter in Illinois was listening. He phoned San Francisco and told us how to turn off the alarm.

During the first month and a half following the earthquake, The Salvation Army helped over 227,000 people, served 101,472 meals, distributed groceries to 65,687 individuals, and gave away 178,825 pieces of clothing, 1,919 pieces of furniture and 27,396 medical items.

In the first two months, the organization distributed $3.6 million in goods and emergency services including $200,000 in rental assistance to San Franciscans and $30,000 worth of building supplies.[40]

The Money Stayed Here

Ultimately, The Salvation Army raised $19 million and filled 19 warehouses with goods in kind following the quake. After months of emergency and intermediate assistance, money was still flowing in from throughout the country. Emergency needs had been met. Intermediate recovery help was given. What would the next step be?

Of monies and goods collected following the earthquake, the Army allocated $15,677,817.71 for transitional and long-term housing facilities and ongoing earthquake-related expenses. These included Harbor House/Gateway for alcoholic single parents with children in San Francisco, a Silvercrest housing facility for seniors displaced from their homes in Santa Cruz and transitional housing facilities in Watsonville and Oakland. Money was put into the emergency assistance arm of the Senior Meals and Services program in San Francisco, whose kitchen was outfitted for major service in emergency conditions. The remainder was spent on ongoing earthquake-related expenses in hard hit areas of Watsonville, Hollister and San Francisco.[41]

Better Prepared

After these years of increased service at disasters in other regions, Loma Prieta made it clear that better relief preparation was needed in San Francisco.

Working in conjunction with the Mayor's Office of Emergency Services and other agencies, the Army improved its local disaster readiness plans, including Bob Moorhead's preparation of a PREPKIT®, an all-inclusive disaster relief guide with magnets to hang on a refrigerator door. It's now distributed nationally.

In local disasters, the Army has emerged as the major food provider. Its spaghetti and meatballs and other full meals often feed fire victims from downtown single room occupancy hotels. Its canteens roll at midnight, whenever the call comes in. If needed, its buildings can be used as shelters.

9/11/01
Eight Months at Ground Zero

Lt. Colonel Ray Robinson addressed earthquake relief efforts in Mexico City. Lt. Colonel Bruce Harvey dealt with El Salvador, Northridge and Loma Prieta earthquakes. By the time Lt. Colonels Richard and Bettie Love took over leadership of the division, The Salvation Army in San Francisco was working hard to strengthen local response capabilities.

Ironically, the biggest catastrophe during their tenure happened on the other side of the country.

It is Salvation Army policy and just good sense that those closest to a disaster are the first to respond. On September 11, within the hour, East Coast Salvationists were at the World Trade Building, the Pentagon and the field near Shanksville, Pennsylvania. The response was overwhelming.

At the same time, a decision was made about asking for funds. The National Travel and Safety Board was involved. People's lives were at stake. Emotions ran high. It was a vulnerable time. At the request of its Western Territorial Headquarters, The Salvation Army did not ask for funds until nearly a week later when it could more accurately assess its long-term role in the relief efforts.

*The Salvation Army is well known for its warehousing capability.
What it receives is quickly distributed. Photo by Kazuhiro Tsuruta*

In San Francisco, the Army was on duty at the Turk Street Command Center within an hour of the attack. The most important thing on San Francisco's agenda was to assure that its own response system was ready, that everything was in place as the nation began to worry about further terrorism. Joye Storey stayed at the center for a week, part of the team who had worked to prepare for any eventuality.

Although local volunteers wanted desperately to go East to help, there were more than enough people from other parts of the country already hard at work. Salvation Army warehouses on the East Coast filled immediately with an overwhelming number of supplies from all over the country. Within a week, Salvationists had to ask for a halt to take time to sort them out.

The Salvation Army's primary place of business at Ground Zero was a tent—actually an aluminum, heated and air-conditioned structure the size of two football fields. Jokingly, but with good reason, they called it the Taj Mahal. Huge semis were parked inside.

This gigantic white structure served as a dining room and rest center throughout the nine months of digging for victims and clean-up that took place at the World Trade Center site.

During that time, the Army provided over 500,000 meals.

When relief workers came off their shift, Salvationists removed their heavy boots, washed and powdered their feet and gave them clean socks. When a body—or body part—was found in the rubble, Salvation Army officers were often asked to pray and accompany the cortege of workers who took it to the morgue.

Major Evelyn Chavez from San Francisco was asked to preside over a ceremony for a body bag that included only a scalp and a fragment of woman's underwear found with it.

This is Not the Red Cross

Their logos are both red. They both respond to disasters. But The Salvation Army and The American Red Cross are not the same organization. Trying not to duplicate services, The Salvation Army often goes out of its way to be in a different place from where the Red Cross has set up camp. Even when it's in the middle of a covey of cameras, the Army is so much part of the disaster relief scene it's simply taken for granted. The logo is red, just like the Red Cross. People often confuse them. And although the Army sometimes arranges to supply food at cost for the Red Cross, at other times it isn't reimbursed and asks the public for donations to cover expenses. A Memo of Understanding describes the role each agency plays. Historically, there is some latitude as decided by local communities. Each situation is different. The Oakland Hills Fire in 1991, for example, was an outstanding cooperative example of sharing the load. The Red Cross worked one side of the freeway. The Army worked the other.

The Long, Hard Clean-up

Long after the TV crews have gone home, after the headlines have moved into the next news cycle, the Army stays in a disaster relief mode. A public service announcement after the Central California floods of 1997 asked the question: "Can you stand the stench of food rotting in a refrigerator after the electricity has been off for weeks? Are you willing to walk ankle deep through the mud and raw sewage? If you have the stomach for it, the backbone and the will, The Salvation Army needs volunteers to give hands-on assistance to people returning to flood-soaked homes."

The message was reminiscent of 1993. Then, in tiny towns along the Mississippi River, Army volunteers spent months helping hundreds of people rebuild homes that had been covered with 49 feet of water for almost a month.

Trying to put his feelings into words, Major David Dahlberg reflected on the long personal ties so many people have with the land and the river, usually poor people, those who live in the lowlands. Returning to a home after the flood, he thought, was something like going to a funeral home to view the body of a loved one.

The city inspector was there. What was the value of the house? Would it be worth rebuilding? Should the family move on? Dahlberg stayed with a family as they listened to their options and said Army volunteers would board the house up until a decision was made. A next-door neighbor cried softly, "I hope your new home has a big kitchen so you can bake those wonderful biscuits."

"Don't bother to look for the big reasons we do this work," said Dahlberg. "It's the little things that matter, the way the presence of God can help people turn corners and get on with their lives and create something new."

Notes

[1] "Denial of Disaster, The Untold Story and Photographs of the San Francisco Earthquake and Fire of 1906" by Gladys Hansen and Emmet Condon.

[2] *Ibid.* p 14

[3] *Ibid.* cover notes

[4] "The General Was a Lady" by Margaret Troutt, p.122

[5] Evangeline Booth's speech at Union Square, *War Cry*, May 5, 1906

[6] *Dallas News*, May 2, 1906

[7] *New York Times*, April 22, 1906

[8] *War Cry*, May 19, 1906

[9] Territorial Secretary's notes during the first week, War Cry, May 19, 1906

[10] *New York American*, Wednesday, April 25, 1906

[11] *Ibid.* p 58

[12] Commander Evangeline Booth speaking of the death of Anna Butler, *War Cry*, May, 1906

[13] "Marching to Glory," Edward McKinley, p 134

[14] Commending Elsie Butler, *War Cry*, May 12, 1906

[15] Letter from Colonel George French, April 24, 1906

[16] Ensign Porter letter, April 1906. The letter suggests cost for repairs might be as high as $3,000. Colonel George French's annual report later indicated that although the buildings were destroyed, the cost was pared to $1,000.

[17] Annual Report by Colonel George French following the Quake. History Room, San Francisco Main Library

[18] From a June 23 *War Cry* report by Salvation Army National Commander Evangeline Booth, after her return from San Francisco, where she laid a wreath on the grave of Anna Alleman Butler.

[19] Doughnut Girl references: "Born to Battle" by Sallie Chesham, 1965; Frederick John feature story, *San Francisco Chronicle*, Nov. 30, 1977; Mildred Hamilton feature story, *San Francisco Examiner*, May 8, 1977; Elizabeth Holland feature, UPI; Letter of endorsement by Margaret Sheldon Manufacturing Company, K C Baking Powder, June 1919; Martin Holmes feature story, *Modesto Bee*, July 2, 1990; *War Cry*, May 19, 1984; Frances Dingman, *New Frontier*, Spring, 1989.

[20] *War Service Herald, Official Gazette of War Service of The Salvation Army*, July, 1919

[21] Captain Young, returning home from World War I, speaking to a crowd of civic leaders and well-wishers. Reported in the *War Service Herald*, July, 1919

[22] *Ibid.*

[23] *War Service Herald*, July 19, 1919, p. 4

[24] *Ibid.* p. 306

[25] This story was remembered in "The War Romance of The Salvation Army" by Grace Livingston Hill in 1919. The young soldier, Joseph, wrote his mom about the "mud, mud, mud up to one's knees," saying how often he thought wistfully of the warm fireplace at home. The conditions he described were horrible, but he made sure to add a note of comfort, "Don't worry, mother, I will be home some day."

[26] Memo written by Lt. Colonel Henry Koerner, November 1986

[27] Ann Landers column, June 11, 1989, and May 1, referring to a March 11 column she had written describing the Red Cross relationship to the British High Command and Secretary of War Henry Stimson's order that Allied forces be required to pay for services at Red Cross clubs. Landers reported that she had received more than 25,000 angry letters regarding this column

[28] Judge Advocate General quoted in "Soldiers Without Swords," p. 167

[29] "My Fifty-Eight Years," by Commissioner Edward Justus Parker, National Commander, USA, An Autobiography, 1943. pp 284-285

[30] Edward McKinley, "Marching to Glory," p 224

[31] Lt. Colonel Henry Koerner memories of Pearl Harbor, the first night

[32] San Francisco Kiwanian newsletter member highlights, January 31, 1995

[33] "My Fifty-Eight Years," p. 295

[34] Interview with Colonel Henry Koerner, 2002

[35] Frances Dingman interview with Colonel Victor Newbould

[36] Interview with Colonel Catherine Sammons Bearchell

[37] "A Galaxy of Glory," history of The Salvation Army in Chinatown

[38] Matthew 10:42

[39] "The House That Juan Built" by Winifred Gearing, 1979, published by The Salvation Army, Atlanta GA

[40] Report Prepared for FEMA

[41] *Ibid.*

THREE

Drug Addiction

"Put Another Nickel on the Drum. Save a Drunken Bum"
From the Very Beginning the Army Worked on Skid Row

As the story goes, the first alcoholic in America that The Salvation Army was able to literally, quite literally, get back on his feet was "Ash Barrel Jimmy." It was 1880. The Army had just arrived from England.

When they found him, he was upside down, drunk in a New York ash barrel with his hair frozen to the bottom.

There were plenty of others like him on the street. The song "Lying in the Gutter" told it all. It was a catchy tune—easy to sing, easy to illustrate how low men and women on Skid Row had fallen. The stink, degradation and absolute hopelessness of a person so deeply into the late stages of alcoholism was a stereotype, but it hit a nerve.

People asked The Salvation Army to help people they themselves stepped across the street to avoid.

"Put another nickel on the drum, save a drunken bum," they sang. A big bass drum was used in the Army's open-air meetings. Eventually, it was turned on its side and passersby were invited to leave donations on top. In those days, a nickel would buy the cost of a bed for a night.

It was another time, another era. Salvationists, like the rest of us, remember it as a time in history that has long since passed.

Actually the Army, itself, wrote many of the early songs about demon rum. Old timers were great about simply giving new words to popular beer hall music. "Why should the Devil have all the best tunes?" they asked. So they recycled popular tunes like "Champagne Charlie is My Name" and "The Marseillaise." In America it was tunes like "Old Folks at Home," "My Old Kentucky Home," "In the Good Old Summertime," "Banks of the Wabash" "Mother Machree," "Yankee Doodle," "Sweet Rosie O'Grady," "My Darling Clementine," "When I Grow Too Old To Dream," "Dashing Through The Snow," "Let Me Call You Sweetheart," "Auld Lang Syne" and even "Sidewalks of New York." You know the tune. Try it with the Army's words.

> Inside, outside, all around the town,
> Our Army helps the helpless,
> Lifts the fallen when they're down.
> Homes have been made happy,
> Heaven has blessed our work
> Over all the Union, right
> From 'Frisco to New York."[1]

Salvationists printed the songs in Army publications and sang them with gusto. It was a common touch, a way to reach out to the crowds with images everybody knew and songs that were easy to sing.

Those old-timers knew how to attract a crowd. The songs were great favorites.

Lying in the Gutter

I was lying in the gutter, all guzzled up with beer,
Pretzels in my whiskers, I knew the end was near,
But a bowl of beans and Jesus' tears they saved me from the hearse.
Glory, Glory Hallelujah, sing the second verse,

Hallelu, Hallelu, put a nickel on the drum,
Save another drunken bum
Hallelu, Hallelu, put a nickel on the drum,
And you'll be saved.

Once when I was young, I was the village belle,
But the way I carried on, I was headed straight for Hell.
I rode my tandem bicycle with my ankles in full view,
But now that I have seen the light, I'm a maiden wrought anew.

Chorus

I was purveyor of fine liquors by appointment of King George,
I posed for Calvert—glass in hand—distinction made me large.
But I took a sip while posing—my descent was swift and hard,
Now I am a common laborer in the vineyards of the Lord.

Well, it's G-L-O-R-Y to know that I'm S-A-V-E-D,
I'm H-A-P-P-Y to be F-R- double –E
F-R- double-E from the wage of S-I-N,
Glory, Glory, Hallelujah, Sing that song again.

Away with Rum, By Gum

My mother said never bake cookies
Because cookies have yeast
And one little bite turns man into a beast
Oh, can you imagine
A sadder disgrace
Than a man in the gutter
With crumbs on his face.
Away, away with Rum, by gum,
With Rum, by gum, with Rum, by gum.
Away, away with Rum, by gum,
The song of The Salvation Army.[2]

Once I was a Water Street bum,
Full all the time with five-cent rum,
The devil had me under his thumb,
When I was in his Army,
Oh! yes he had me in his grip.
I cared for nothing but a nip,
I down the road to hell did slip
In the damnation Army.

Glossary From the Depression Era

During the Depression, men came west looking for jobs. Unfortunately, a lot of them found the bottle instead. Young Ardys Smith was a captain at San Francisco's Harbor Light years later when she was part of a presentation describing how often men talked about being separated from their families. What songs did they want to hear most? Those about "Mother," of course.

In the same program, Captain Stan Davey described images of a long ago past, including stand-in line street meals and sermons-before-you-eat at rescue missions like the Army and several others. The Army wasn't the only mission on Skid Row, but it was part of a street culture that had a vocabulary of its own. The definitions were very graphic descriptions. Captain Davey—known for his progressive program changes—saved the old-fashioned phrases for the archives.

Ear-Banging	The message by the evangelist before you get to eat.
Knee-Bend or **Nose Dive**	Going forward to the altar for more than the meal, usually for a bed.
Flop	The bed given out at the rescue missions
Sally	Harbor Light or any facility run by The Salvation Army for Skid Row men
Hymns	Singing for your supper (Most asked for hymns: "The Old Rugged Cross," "In the Garden," "Amazing Grace," "What a Friend We Have in Jesus," and "Onward, Christian Soldiers.")
Gourmet	Any food other than beans, needing a fork to eat (turkey neck stew, meat stew vegetables)
Shop Around	Check out missions for shortest sermons, best food and beds.
Holy Rollers	Where they have a lot of "Amens, Praise the Lord and Hallelujah" from the people doing the service.
Change the Sheets	For the all-night movie sleeper when they changed the three or four films shown nightly.
Mick	Smallest bottle of wine
Security	A small jug to get you through the night.
Insurance	The big jug for getting through the night and to have an eye-opener in the morning.
Wine Bells	The bells the Catholic church rings for 6:00 AM Mass, the same time the wine stores open.
Shake Room	Old time detox, usually just a room with a mattress on the floor where you went to get over the shakes.
The Man	Beat Cop who walked Skid Row and picked up drunks
The Wagon	The paddy wagon you were hauled to the drunk tank in (also periods that you tried to quit on your own)
To Stem	To walk down the street and hit people up for enough to get a "small one."
Okie Suitcase	Shopping bag for your personal effects
Okie Pullman	Boxcar
Slave Market	Casual labor, where you worked daily a different job and received vouchers cashed at a nearby bar
Spider	The small amount of liquid left in a bottle, combined with other spiders would get you a drink to get started in the morning[3]

Dorothea Lange, famous for her pictures of the Depression in the California Salinas Valley, spent time with a small but determined band of Salvation Army soldiers in the South of Market, April, 1939. Photos on these two pages courtesy of the Farm Security Administration Collection, "American Memory" Library of Congress

Faith and Works

"Faith and works should travel side by side, step answering to step, like the legs of men walking. First faith, and then works, and then faith again, and then works again."
—*William Booth, in a letter to Salvationists, Sunday, May 30, 1908*

Seen through a neighboring hotel, The Salvation Army mission provided assistance in a forgotten alley on Skid Row.

Here's My $5

"Had eye surgery this year, cataracts. My leg is getting worse. Will probably have it amputated. Lots of doctor bills. The roof leaks. Times are rough. But here's my $5. Make sure it gets to somebody who needs it."

—*Letter from a donor during the 1980's*

DRUG ADDICTION

"Falling Off the Wagon"
The Phrase Started with a Real Wagon
And an Annual Parade

Unfortunately, when a hard drinker "falls off the wagon," it usually happens with a resounding thud.

No matter how many days, or months or even years a person's been sober, one drink can send the alcoholic demons reeling again. To be "on the wagon" is a sign of recovery. To be off is to be caught up in the revolving door of addiction that often gets far worse before it ever gets better.

Most people know the phrase. Few know it was a promotional stunt by The Salvation Army that gave it legs.

There is a theory that the phrase may have started in England and had to do with a prisoner being allowed one last drink on his way to the gallows. In America, however, it started with a real wagon and real alcoholics—who stayed on board, or fell off depending on how hungry they were for a good Thanksgiving meal.

In 1909 in New York City, the Army's National Commander Evangeline Booth presided over her first "'Boozers' Day Convention."

Like her father, Evangeline understood how to get people's attention.

Her event was so outrageous, so unique and evidently so popular that it continued every year until Prohibition in 1920.

"Free eats all day," said signs posted on flophouse fences. For weeks in advance, billboards and posters announced a parade with a big rally at the end. A fleet of borrowed Fifth Avenue coaches cruised around the Bowery, public parks, flophouses and Hell's Kitchen. Anybody in need of a meal could hop on board. As many as four or five marching bands played. People joined the procession all along the way.

In the middle of it all, loaned by the city's street cleaning department, was a water wagon drawn by four horses. Boozers climbed on. They were on their way to get something to eat and dry out.

If they "fell off the wagon," it was assumed they were going back to the sauce.

Newspaper editors seemed to love Boozers' Day. One year, a clipping service said thirty-six yards of single-column space was given to the event by New York newspapers alone. Henry F. Milans, one-time managing editor of the *New York Daily Mercury*, took it seriously. A well-known alcoholic, he joined the parade and, when last heard from 19 years later, was still sober.

They say the first Boozers' Day parade included a walking whiskey bottle ten feet high and floats dramatizing scenes of a drunkard's life.[4]

Talking to the Guys
At a Fifth and Stevenson Saloon, 1935

"One Sunday afternoon, twenty-three years ago, I was strolling leisurely up San Francisco's famous Market Street. As I came to the corner of Fifth I noticed a small crowd standing near the curb. Drawing closer I saw that a number of Salvationists were holding an 'open-air meeting.' One of these 'strange' people was standing on a chair, talking to the crowd.

"I stopped and listened to him awhile. After a few moments I threw a half-dollar on the drum and walked on to join a few of my old friends in a saloon near Fifth and Stevenson Streets.

"That Salvationist's talk, however, had aroused my interest and after that night I always managed to be at Fifth and Market when The Salvation Army was holding its 'open-air meetings.'

"One Saturday evening, a few weeks later, I was with my 'pals' in the saloon, and had a glass of beer in my hand. Suddenly The Salvation Army went marching by . . . 'Boys,' I said joshingly, 'this is my last glass . . . I'm going to join The Salvation Army.' How true that was going to be . . . Before I left The Salvation Army hall that night it was no josh . . . it was true! I knelt at the 'penitent-form' and got converted. Since that night I haven't touched a drop of liquor or gambled a cent.

"Am I sorry? Don't I wish I could go out and hit the 'high spots' and have some 'fun'?

"No, I'm not sorry in the least. I'm having a better and happier time now, working for God, in the Army than I could have had in a lifetime of 'hitting the high spots.'

"Why don't you try it? You won't be sorry!"

—Charles Sholin
Sergeant-Major S.F. #2 Corps
55 Lapidge Street, San Francisco, California
Social Service Plus"
Territorial Annual Report, 1935

From Skid Row to Suburbia

Skid Row was the popular name given to the seediest part of town where the Army and other missions worked. Back in Seattle, where the name first originated, it was Skid *Road*, the planked road loggers used to roll fresh cut timber from the forests down to the water. In San Francisco, Mission Street was the old plank road, stretching from Mission Dolores to the Bay. And where was Skid Row? Right off Mission Street, of course.

To be "on the skids" is to be down on your luck in the worst part of town.

People didn't talk much about alcoholic women in those days. Mostly, the assumption was that alcoholics were surly men who spent their days guzzling wine or, if they were married and even working, went home to beat their wife. Even that gave rise to verse:

Jim Jones had a brute of a temper
It caused lots of trouble and strife.
When someone downtown would not please him.
He'd go home and thrash his poor wife.

But the Mrs. is now very happy,
And the kiddies shout, "Hip, hip, hooray!"
For instead of his wife, now he's beating
The drum in the Army today.[5]

The songs seem quaint, the way old time ditties from the early part of the century do. For years people sang the choruses on the street corner, around the fireplace, on shipboard with the crew during wartime, at college fraternity parties. Many of the songs became part of folk music history.

But as addiction became more prevalent, as people realized the same problems were happening in middle class and wealthy homes, it wasn't quite so easy to assume a drunken bum was somebody else's father, somebody else's son. Ultimately, the caricatures didn't seem funny anymore.

It didn't happen immediately. In the 1950s there were still popular images of the tipsy suburban husband at a New Year's Eve party in his ranch style home wearing a lampshade and tripping over furniture. During his "Rat Pack" days, Dean Martin often had a glass in his hand as he performed on stage. Was it liquor? Was it apple juice? It didn't matter. Even if it was an act, the audience loved it. He could slur his words, but his voice was golden and he always got a laugh.

Today nobody laughs at a falling down drunk. Stand-up comedy routines from the 1950s and 1960s seem painfully demeaning in a politically correct, medically advanced, more sensitive world. But today when someone falls off the wagon—assuming they ever got on it in the first place—there's still a thunderous thud!

Please, Please Help My Son

No matter how different each case of substance abuse is, the story is often the same. Someone's life is out of control.

"Please, please," begs a parent, "give my daughter a place to pull things together. Help her believe things won't always be so awful as they are right now."

The answer is gentle and non-judgmental. Facts. Resources. Referrals. Step by step advice. But there's an underlying truth. It's not enough for friends and family to want help. The addict has to want it enough to walk in the door and ask. Want it enough to say, "Here I am. I am helpless to do it myself. I need help."

There's a basic premise here. The Salvation Army doesn't change lives. It provides a venue for people to change their own circumstances. Addicts are responsible for their own recovery. The Army provides the place, the program, the hope.

On the road in San Francisco. Each year, as access to drugs becomes easier, the Army works with young people who have become alienated from family and friends. Photo by Judy Vaughn

Harbor Light is a recovery community. By its very nature, this sets it apart. But, as much as possible, counselors try to remind clients this is the real world—not an insulated society away from the pressures of ordinary day-to-day living.

Residents are encouraged to find work, put their medical and legal affairs in order, re-relate to family from which they may have been alienated and—in a peer setting—learn to take responsibility for their own recovery.

Just a few years ago, a large proportion of Harbor Light and Harbor House clients in San Francisco had high school diplomas and a semblance of job histories. Today, most haven't finished high school. Since drugs are so easy to get at an early age, many have been addicted longer and have fewer job skills and longer experience on the streets. Substance abusers often come with multiple diagnoses—alcohol, hard drugs, AIDS, bouts with depression, periods of homelessness, severe alienation,

"To some," says a staff worker, "this is the first real home they've had in years." Although the program is spiritually grounded, church attendance isn't mandatory. For those who choose, a "Celebrating Recovery" program on Friday nights combines a three-hour opportunity for fellowship, family entertainment and intensive group therapy.

Does anybody ever get better? Is there hope for even the most heartbreaking cases? Ask people who've been through the worst. Ask them to remember what it was like. Ask their families how they lived through the hard times.

Then ask graduates of the program who, years later, are still counting anniversaries of being clean and sober. Ask them if it was worth it.

Detox

From the street, you can't see The Salvation Army's Harbor Light detox facility on McLea Court in San Francisco. But the Mobile Assistance Patrol knows where it is. Doctors in the emergency room at San Francisco General know. The cops know. Desperate families of substance abusers know.

Eventually, some of San Francisco's most tortured alcoholics find it.

Inside, there are 24 beds, half of which are reserved for referrals from Swords to Plowshares and the Veterans Administration.

The rest are open for the individual who calls on his

or her own to say, "This is it. This drug is killing me. I don't want to do it anymore. I need help."

Walking in the door is hard. People in charge of the program paint a famous but not so unusual picture of a wife bringing her husband in for detox. The farewell is tearful. "Please get better. Please get better," she begs. "This time I will. This time I will . . ." is his reply. Then the client says goodbye, walks into detox and straight out the back gate. The next year it happens again . . . same couple, same scenario.

Considering that these are clients who've mostly conned their way through years of hard drinking and using, there is an honesty that seems genuine among some of those who stay.

"Oh, we've all got plenty of guilt," says one. "I once sold my kids' Christmas tree and all their presents for drugs. I got about $100. On Christmas Eve, they had presents under the tree. Christmas Day they were gone. I guess that's about the worst thing I ever did."

She remembers the horrible hours just prior to coming to detox. "There I was sitting on a crate for five days in front of St. Anthony's, barefoot with a blanket around me. I had smoked up $800 worth of drugs and I was stretched out on the sidewalk just like any other bum. I used to frown on people like that. Now it was them frowning on me.

"And it was my birthday! It's a bad thing to be tore up on your birthday. A nun at St. Anthony's gave me $10 so I could admit myself to Harbor Light." (This is a nominal fee, giving clients a sense of investment in their decision to enter—the decision to spend the money on recovery, not another drink.)

She hasn't a single tooth in her mouth, her hair is unruly, and her body is hopelessly out of shape. She says she's been to 20 different treatment centers but admits that her recovery was so tenuous in one that she celebrated her graduation with a bottle of champagne.

"I've tried to stop drinking 101 ways that didn't work," says another woman. "I can't believe what I was doing. I was sitting in a rat-infested crack house for two days, a dirty, *nasty* house. There were rats in that house! Then, when I started seeing it for what it was, I was so paranoid that I jumped in a cab at 6:30 in the morning to come here . . . When I started telling the truth, it got easier."

This is a young woman on leave from a good job to address her issues. She looks in the Alcoholics Anonymous Big Book again, reading quietly, setting herself somewhat apart from the group but, from her actions, making it clear that she knows recovery is a long, long road.

On the nights before he entered detox, a man was sleeping on the ground near a train track. He remembers looking enviously at people on the train. "Why should they be on the train and not me?" he asked himself. "Why wasn't I on that train?"

Why indeed? He presents himself as an ideal candidate for a job—clean-cut, soft spoken, well-mannered. Yet he has inner demons that manifested themselves when he was a child and multiplied as an adult.

"Don't worry," he says, "I can always get a job. I've had about 50 in five years." In the Air Force, he had originally wanted to be a pilot, but didn't qualify since he hadn't finished college. The last job he walked away from was far from the dreams of his youth. He was a truck driver.

Other kids picked on him as a child, but he didn't defend himself, didn't fight back. So his anger was bottled up and when, as an adult, he finally began to let it out, he had an attitude that wouldn't stop.

"Anybody who threatened me was my enemy. I was full of rage. Wouldn't even salute officers in the military. I wanted respect. I thought I had earned respect.

"But in the end, I felt I wasn't good enough for anybody . . ."

Another man's memories of recent weeks, after eight years clean and sober, seem to concentrate on the Tenderloin (where it was safer to be homeless, he says, because of the bright lights) and Sixth Street, a place that figured heavily in his addiction years ago. How he ended up there again after so many good years weighs heavily on him. He's emotionally drained.

"I'm not dumb," he says. "I just do dumb things."

"Stay off Sixth Street," he repeats. "I used up $2000 in two days." He hit bottom when he suddenly remembered The Salvation Army (known on the street as "Sally") where he had detoxed eight years ago.

"Thank God," he says, "Sally still is here."

"Whatever else I know about The Salvation Army, I know it was there for my son when he needed it. It's a wonderful organization."
—*Grateful parent*

Often Inherited, Always Developed by Indulgence, But Clearly a Disease

"After a time the longing for drink becomes a mania. Life seems as insupportable without alcohol as without food. It is a disease often inherited, always developed by indulgence, but as clearly a disease as ophthalmia or stone."
—"In Darkest England and the Way Out," 1890

As early as 1890, William Booth suggested the road to alcoholism had a medical foundation. Developed by indulgence, it had the potential to become an addiction. And indulgence sometimes was simply following the path of least resistance.

Booth wrote of the appallingly overcrowded housing of London's slums and came to the conclusion that the local saloon not only was a place for a person to drown his sorrows, it was also a kind of community center. If misery loved company, and it certainly seemed to, then the place to go for solace was the local pub.

"The taproom in many cases is the poor man's only parlor," said Booth. "Many a man takes to beer, not from the love of beer, but from a natural craving for the light, warmth, company and comfort which is thrown in along with the beer, and which he cannot get excepting by buying beer. Reformers will never get rid of the drink shop until they can outbid it in the subsidiary attractions which it offers to its customers."

"Let us never forget," he warned, "that the temptation to drink is strongest when want is sharpest and misery the most acute . . . Gin is the only Lethe of the miserable. The foul and poisoned air of the dens in which thousands live predisposes to a longing for stimulant. Fresh air, with its oxygen and its ozone, being lacking, a man supplies the want with spirit."[6]

"I must assert unhesitatingly that anything which dehumanizes the individual, anything which treats a man as if he were only a number of a series or a cog in a wheel, without any regard to the character, the aspirations, the temptations and the idiosyncrasies of the man, must utterly fail as a remedial agency."[7]

Harbor Light History by the Year, How It's Changed

The Salvation Army has always been a presence in the South of Market. As early as 1885 on Skid Road, it provided services to indigents and alcoholics at 47 Hunt Street (today a tiny unmarked alley between Howard and Mission next to the Museum of Modern Art). But it wasn't until 1941 that it adopted the name "Harbor Light" and called it a "place of new beginnings."

By 1950 the Hunt Street building was already too small. In 1953 District Attorney Thomas C. Lynch called for a complete reappraisal of the way San Francisco was meeting its alcoholic problem. At a Salvation Army luncheon, he chided the City for not following up on a 1948 Daly Report to deal with what he saw was not a criminal problem . . . but a social problem, a medical problem. Lynch reported that of San Francisco's 40,000 misdemeanor cases in 1952, a great bulk of them were directly attributable to drinking. He felt the large percentage of alcoholics could "be taken completely out of the criminal picture and rehabilitated."[8]

The Salvation Army accepted the challenge and ultimately moved its soup line and hospitality house (20 beds) to 240 Fourth Street. Jack Block was an early volunteer. Years later, when he reminisced about the old days, he said San Francisco business people were eager to participate. "We would go to a meeting and say we'll need such and such and a dozen hands would go up."

In 1953 a *San Francisco Call Bulletin* editorial applauded the Army's work, but moaned that San Francisco had every reason to be ashamed of its Skid Road. Quickly it noted that "we're by no means the only city smirched by such conditions. But . . . we can be proud of such work as is being done to bring the light of hope to this latter-day Slough of Despond."[9]

In 1956 the Army acquired an adjacent hotel with 30 rooms and named it James House for James Dinwiddie, a major supporter. According to the *San Francisco Examiner*, the organization paid $50,000 for the Avondale Hotel at Fourth and Howard and was spending $75,000 to renovate it (a shelter for 40, dormitory for 30, hotel housing for 45.) Members of the Harbor Light Citizens Committee headed by Dinwiddie worked three years on the project. Captain Harold Ferguson, Brigadier Cyril Clitheroe and then Major Ferguson served during those years.

In 1966 the Southern Police Station was remodeled to become Bridgeway House, a jail-release program for substance abusers from San Bruno. When the Redevelopment Agency moved to demolish the building, men were housed at Pinehurst Lodge in the Avenues, but "Not in My Back Yard" (NIMBY) neighbors were opposed. Detox services continued temporarily on the sixth floor of the Mars Hotel at the corner of Fourth and Howard.

In 1972, Harbor Light moved into the large Gladding McBean building at 1275 Harrison. In 1974 it was State-licensed with a social model detox, men's residence and the City's first women's recovery program.

Soon after, the Bridgeway Project "sober hotel" opened at St. Luke's Hotel at 242 Turk in the Tenderloin. The Turk Street Corps was established on the ground floor and in 1983 an emergency shelter (later called "Lifeboat Lodge") opened from the other side of the building at 341 Eddy Street. Captain Stan Davey served from 1961 until 1976, followed by Major Les Sparks (1976-1981); Captains David and Effie Patrick came next (1981-1984) with A/Captains Nick and Ruth Gabriel following in June of that year, along with Captain Donna Bales. Majors Larry and Vickie Shiroma were in charge until 1999, with people like Captain Richard Reuer assisting. They were followed by Captains Phil and Lowry Smith.

When an Alcoholic Has AIDS

In 1990 The Salvation Army was one of the first agencies in San Francisco to seek funding for AIDS treatment. These were the early days. The public was only just beginning to understand the impact of the AIDS crisis, but The Salvation Army was at the forefront of dealing with specialized programming in the recovery community. Today, on any given day as many as a dozen of the program's current clients are HIV-infected.

In the mid to late 1980s, what had previously often been described as an alcohol and mental "dual diagnosis" progressively became a "triple diagnosis." It soon became apparent that a number of clients suffering from addiction and depression in San Francisco also had AIDS. Harbor Light observed universal health precautions, added specialized supportive services, developed a strong education program and started looking for new monies to respond to the growing problem.

When Federal Ryan White Care Act money became available through the City in August, 1990, the Army applied immediately. Funds were awarded in 1991.

"It was a very emotional time," says Byron Hudson, a former employee. There were people succeeding in treatment suddenly jolted into the reality of a life-threatening disease about which there was little information. Others, already diagnosed as HIV positive and afraid for their lives, felt helpless about attempting any kind of substance abuse recovery. It was a time of increased awareness of disability needs, increased need for specialized staffing.

"Whatever the city saw as a need, whatever they asked for in treating substance abuse—assistance for HIV clients, deaf clients, wheelchair clients—The Salvation Army came to the call," Hudson remembers. "We tried our very best."

Carrie Graham, MSW, MPG, worked there as Harbor Light's Health Services Coordinator from 1992 to 1994. She remembers describing the Army's program to a number of community groups, including the Gay/Lesbian/Bisexual/Transgender Task Force, Disability Task Force and Baker Place.

"I think they may have felt there was a perception in the community that The Salvation Army might have difficulty working with the HIV-infected gay community. But after a few referrals they found out it just wasn't so.

"By our actions and presence, we were raising awareness. Thanks to the openness of Larry and Vickie Shiroma, we made great strides."

Graham left Harbor Light after two years to continue her schooling. In 1999 she returned to write grants and helped facilitate a program with Kaiser, which does not have residential services, but contracts out with recovery programs and hospitals for certain clients in their outpatient program. It's a per diem arrangement. Harbor Light staff does the work. Kaiser pays the bills.

Joe Gonzalez, L.C.S.W. CDRP, Program Director of Kaiser Permanente Medical Group, Inc. in his support for the program, noted that Harbor Light plays an

> 1974—Harbor Light Offered the City's First Women's Recovery Program
>
> 1990—Harbor House Offered the First Licensed Day Care Program in Transitional Housing for the Homeless
>
> 1990—Harbor Light was one of the first two agencies in the city to apply for and receive Ryan White CARE Act funding to treat substance abusers living with HIV/AIDS.

important role in the continuum of service. There are strong success stories among the patients referred to Harbor Light, a number of whom are HIV-infected. "They've used their treatment at Harbor Light to build a solid foundation for ongoing recovery and have subsequently returned to Kaiser to make good use of the outpatient program."

The original 10-bed integrated recovery program in 1991 provided detox, primary and recovery program activities, specialized counseling and support groups, plus linkage to community health care facilities.

In 1994, thanks to a second round of Ryan White Care Act funding, services were expanded still further through an additional 12-bed, 21-day co-ed facility for individuals with HIV/AIDS. Services include specialized counseling and community linkages including transportation for emergencies and primary care appointments. Officers and program staff work closely with community resources in the field of chemical dependency, especially with agencies such as McMillan House and Friendship House, Kaiser Medical Center and San Francisco General Hospital.

In 1998, $1,164,000 in annual funding was lost when the Army did not offer employee benefits as described by Domestic Partners legislation. It was an extraordinarily sad, difficult, polarizing time. Private donations kept the program open, but at a reduced level.

The work continues. And, unfortunately, so does AIDS. As a resurgence of the disease continues in both the gay and straight community, Harbor Light continues to see clients who are affected—people with dual and triple diagnoses, people who have been suffering with health problems for a long time.

The story of a client who died of AIDS-related complications at San Francisco General illustrates how raw emotions can be for people in recovery. The man had lived at Harbor Light for some time and his death left the house clearly devastated. Although pain is a daily occurrence for people in the recovery community, this was a particularly heavy loss. Captain Phil Smith, Director, gave the eulogy in a room packed with recovering alcoholics. "We may be beat up and torn up, but we're not shut up," he said. "I've never been in a place where people held each other up so well."

He Lived Two Years in a Bush

"I would like to thank The Salvation Army for turning things around."
—*Thanksgiving tribute from a Harbor Light resident.*
Before entering the program, he had lived for two years in a bush near Civic Center.

Moving Stuff!
The Truck You Call Has More Than One Purpose

You know the truck. In the old days, when it was horse drawn, it was green. Then red. In recent years it was blue, with a cartoon figure reaching for a telephone. In 2004, it was shining white with full color photos of things you'd be proud to donate. Each truck is a moving billboard featuring a bright red shield. People say it's one of the most recognized logos in the world. In San Francisco there are 23 Salvation Army trucks moving your Cousin Evie's old sofa . . . shifting supplies from one warehouse to another during a disaster and—in the age of recycling, volatile dot-coms, corporate closings, consignment shops and sidewalk sales—reminding people that the Army needs reusable goods, not junk. Don't mistake the service for a dumpster.

Most people know the shield. They know the Army picks up, recycles and sells second hand material ranging from old clothes to old cars. What they don't know is that the people doing the recycling are alcoholics in recovery. The sale of every piece of furniture, every piece of clothing and every knickknack helps support a residential work therapy program.

In the early years the program was called the Dusty (short for Industrial Home and, some say, easy to remember because throwaways were always so dusty!). For years, it was the Men's Social Service Department. Today, it's called the Adult Rehabilitation Program (ARC).

Do Something!

In 1887 when the London Bridge still crossed the Thames and not the Arizona desert, Salvation Army founder William Booth was appalled to realize that homeless people were sleeping under London's bridges—a mass of humanity living most inhumanely. "Did you know that was happening?" he asked his oldest son. Sheepishly, Bramwell had to admit he did.

"Well, do something about it!" said the father to the son in a tone of voice we've come to believe must have been pretty emphatic.

And Bramwell Did

On February 21, 1888, the Army opened its first homeless shelter, a warm warehouse with beds and cheap food at 21 West India Road, Limehouse. It wasn't free. In Booth's mind that would be demoralizing. Instead, there was a small fee and, if that was unavailable, there was a chance to work—either with the Army or with outside employers who gave jobs to day laborers. From the very beginning, even when its only purpose was to preach the gospel, The Salvation Army understood that its primary outreach was to the poorest of the poor. When all else failed, the Army was a place where a person could go.

This soup kitchen "pick-them-up-from-the-street-and-give-them-a-bed" philosophy quickly evolved into programs that provided a way for "down and outers" to keep a shred of dignity until they got on their feet again. The Army handed out chits for homeless people and gave them jobs cleaning city streets and chopping wood. In San Francisco, Mayor Adolph Sutro was a major supporter.

The Lifeboat

The same Captain Joseph McFee who took the "Theodora" cruise off to the hinterlands to raise money for social services in San Francisco and who put out the world's first Salvation Army Christmas kettle, was quick to see the future in General William Booth's plan to put men to work.

He was a moving force behind The Lifeboat, which included a "Pill Box" (clinic) and an "Old Curiosity Shop" (thrift store) plus a workshop and Labor Bureau on Grant Avenue.

The key to its success was captured in a word the Army has since made its trademark—recycling. That which others threw away, The Salvation Army tried to save and use another way. Men salvaged rope cables from ships docked at the Embarcadero, then shredded and wove them into mats or used them for upholstering furniture. Grinding up cork bottle stoppers, they made government-approved life preservers and ship fenders.

On the Embarcadero, the Lifeboat's "crew" operated a free reading room with literature in all languages, took care of prison work and prisoners, and even operated a barbershop. Shaves were a nickel. Haircuts were a dime.

On October 3, 1894, an improved Lifeboat opened at 117 Jackson Street. Beds were made of "strong canvas stitched between iron bars, double-decker style." A man could get a bath and breakfast for 15 cents. Additional meals were 5 cents each.

By the turn of the century there were three new facilities for working men—the Working Man's Institute, the Metropole and the Beacon. In 1904, the Industrial was at 271-275 Natoma Street. A 1906 San Francisco directory lists the New Metropole at 147 Natoma, Helping Hand Hotel at 677 Commercial, and Industrial Home and Yard at 275 Natoma.

Property at 1500 Valencia Street was purchased in 1946 and the Men's Residence at 3550 Army Street was added in the 1950s. [10]

Recycling Before it Was Green

Recycling was high on William Booth's agenda long before it was fashionable. Just as he was unwilling to give up on people whose lives seemed no longer useful, he thought discarding old newspapers and magazines was a terrible waste.

Ultimately, wood yards outlived their use, so San Francisco's was remodeled to accommodate the paper industry. White paper, with the manila, was sorted and baled separately. Folded newspapers were tied in bundles and sold to furniture packers. Magazines were sold by the 1000 lots weekly.

Paper came in off the wagons and was hoisted to the loft, where men on either side of a screen table worked it down between them to the chute, which emptied into the press below.

The press was a Salvationist's invention and turned out bales weighing over 400 pounds apiece. The output of paper for the first month (August) was 12,440 pounds and brought $31.10. In December, they sent off 31,310 pounds, selling for $81.30.[11]

When Mom's An Alcoholic

The basis of any Adult Rehabilitation Center program is work therapy. The women's program in San Francisco includes an added incentive—the opportunity to live in a fine old home in the Avenues.

Originally called Pinehaven, The Salvation Army's women's residence now known as Pinehurst Lodge was opened by the San Francisco Junior League in January, 1929. Its purpose was to serve underprivileged children and orphans. By 1935 it had separated from the League and had its own board of directors. The San Francisco Nursery for Homeless Children bought it in 1939. On August 6, 1946, ownership of the whole property was transferred to The Salvation Army as the Pinehurst Emergency Lodge.

During World War II and through the Korean Conflict, servicemen's wives and their children found refuge here, then single women waiting the birth of a child. For a time, the facility was underused. Just as the Army considered selling the property in 1974, Major George Duplain had a better idea.

In the 1970s who paid attention to the middle or upper class mom drinking at home while Dad and the kids were away during the day?

Not many people.

When he set up the Pinehurst program in 1975, Major Duplain was eager to correct the image of The Salvation Army working on Skid Row only. Pinehurst was to be a model for women's recovery programs.

It was hard for the public to accept the fact that women were alcoholics, too. Yes, there was help for women on the street. But those who drank at home, those who hid their liquor from the household and, after their children were gone, felt that life was passing them by, weren't eligible for assistance.

Duplain determined to take the elegant two-story home originally owned by the Junior League and turn it into a treatment center for women. They worked in the thrift store recycling center during the day and had access to a full counseling program. And at night they went home to extra amenities—a piano, fireplace, hot tub and sauna.[12] For follow-through, he also opened a popular re-parenting program at Lytton Springs in Sonoma County to give women in recovery quality time with their kids on weekends.

TOP AND BOTTOM: ***The Salvation Army's Pinehurst Residence in San Francisco
is a quiet home for women recovering from drug addiction.***
Photo courtesy of the Adult Rehabilitation Program

When Grown Men Cry

Casually, very casually, the man sitting in the second row from the front rubbed his eyes. Then his cheek. Then his eyes again.

He looked straight ahead, only occasionally reaching up to touch his face. You could see it was a tentative gesture. It said grown men don't let other drug addicts see your emotion. Grown men don't cry listening to a chorus of pre-teens singing on stage. He was obviously moved. Possibly the father of a teenager of his own.

The songs said, "Things change. Plans fail. Storms rise. The ride's too rough . . ."

Throughout the performance, he periodically wiped away tears.

He thought no one was watching.

The Salvation Army chapel was packed. You could feel the energy. You could easily see people had survived enormous pain to reach this point. Many of them—dually and sometimes triply addicted—had been strung out on alcohol and other drugs since they were teens.

Most were clean-cut young men. Not the grizzled, tired-out Skid Row bum of the old days traveling from one Salvation Army program to another. Not the displaced patient after the closing of California mental institutions in the 1970s. Not disillusioned Viet Nam vets returning from war with a history of drugs. Instead, they seemed more like somebody's kid brother dressed up for church.

To see them in clean shirts and slacks, some in suits, you wouldn't have known that just a few months ago, most were either in jail or on the street. Or perhaps stealing from their families to support their habit.

All day, they had recycled clothing, furniture and goods other people call trash. Some of them were assistants on Salvation Army trucks picking up recyclables throughout the city. Some worked in the sorting room, some on the loading dock. Some in the kitchen. Some were in charge of baling and compacting rags for recycled industrial use.

This evening after work, they had gone back into the residence and cleaned up for the monthly graduation service.

Two people were graduating. A third, on her second attempt, had made the fatal mistake of going home for the holiday weekend. When she returned, her UA (urine analysis) tested dirty for cocaine. Several people spoke about it. Everybody knew. There was an undercurrent of sadness. She was so close, but once again had failed. For her, there would be no graduation tonight. Instead, she faced another round of coping with her addiction alone. Until she could be reinstated, she would be doing it without the six-month residential Adult Rehabilitation Center program in which the Army combines work therapy, counseling and a strong social model of group therapy with peers.

Others will take her place. Of the 18 million people with alcohol problems in this country, and the additional 5 to 6 million with other drug problems,[13] there is always a next person in line. Most are referred through the courts.

"Good evening, family," said one mentor after another introducing new people to the group, some who had tried the program before, some who had only just begun to consider the concept of sobriety.

"There are two kinds of people, who come to this place," said Phil Beebe, long-time program consultant. "Those who do what they have to so they can get by . . . and those who see this as the best six months of their lives, a time to learn how to live without drugs and alcohol."

Some in the room had accepted the fact that their lives had become unmanageable, that their addiction was more powerful than they. Some seemed not quite ready to join in.

Clearly others were. As a young woman stood to accept her diploma, she was greeted with great applause from her peers and Women's Auxiliary members who had come to cheer her on. Her valedictory statement was a clear testimonial to the friendships she had made.

Thank you, thank you, thank you, she said to the Army and to one after another of her friends. She looked each of them in the eye and called each of them by name.

The second graduate elicited an even more thunderous ovation. Since there were more men than women in the program, their support of their buddy was filled with deep bass voices, foot stomping and the kind of cheerleading you would expect from a college fraternity as their team comes out on the field.

A shy man probably in his early thirties, the graduate was dressed in suit and tie. He took his place at the podium, smiled gently at the guys in the back row and tried to put his gratitude into words. He introduced his parents and sisters and apologized to them publicly. "I'm sorry what I put you through. You didn't deserve that. I'm sorry what I put you through. Thank God for places like this where you can learn to control your anger. "

Like the man in the second row, the parents tried to hide their tears.

And the audience applauded wildly.

There was a shared spirit of success in the room. Nobody said beating addiction is easy. It's not. As a mother at one of these graduations once said, "Thank you, thank you for helping our son. He's been through ten different programs. The Salvation Army was our last hope."

Unfortunately, many mothers' sons, and daughters, and husbands don't succeed. Many go through years of trying, failing, trying again.

But in this moment, on this night and in this place, there was a feeling that regeneration really could happen. Salvationists believe in miracles.

You probably already knew that.

They don't give up on anybody.

Notes

[1] "Songs of Yesteryear" from the Evie Dawe collection

[2] Discussion of Kansas History, Kansas-1@listproc.cc.ukans.edu

[3] Glossary included in "How it Was, What Happened and What it is Like Now." a report made by Captain Stan Davey, Ph.D, in 1985.

[4] "The General Was a Lady," p 128

[5] "Combat Songs of The Salvation Army." Sallie Chesham, editor.

[6] "In Darkest England and the Way Out" p 56

[7] Ibid, p 80

[8] *San Francisco Chronicle*, June 23, 1953

[9] *San Francisco Chronicle*, September 26, 1956

[10] "Through the Years," undated document prepared by the San Francisco ARC.

[11] *War Cry*, March 5, 1904

[12] Major George Duplain interviewed by Ezekiel Green for "California Living Magazine," *SF Sunday Examiner and Chronicle*, May 4, 1980

[13] The National Council on Alcoholism and Drug Dependence

FOUR

Changing Times

*101 Valencia Was the Soul
Of the Army's Administration, Worship, and Training*

If there's one building in San Francisco that epitomizes the early history, administration, people and soul of The Salvation Army in this City, it's 101 Valencia. Even though it was sold in 1989 and the offices have long since been converted into condominiums, the historic San Francisco Citadel sign next to the front door says the Army will always be part of this place.

To Salvationists, the building is more than a San Francisco landmark.

It was home base for 69 years.

Commissioner Adam Gifford opened the Western Territory in 1920. From then until 1976 when Territorial Headquarters was moved to Southern California, all officers in the western United States reported here.

From 1920 until 1928 and again from 1933 to 1948, future Salvation Army officers studied in the training school and lived in the dormitories on the top floors.

From 1976 to 1989, all officers in the Northern California and Nevada Division reported here.

And during all those years, 1920 to 1989, the downstairs auditorium was the Army's primary center of worship in this area, the San Francisco Citadel—location of special events, marriages, funerals, Home League bazaars and weekly services. Even though it eventually moved to 19th Avenue and then to South San Francisco, the Citadel will always be the direct descendent of Corps #1 opened by Major Wells on July 21, 1883.

Kelso, Newbould, Goldthwaite, Angel, Taylor, Adams, Wiseman, McAbee, Pack, Covert, Rocheleau and Yardley—there are many familiar names on the roster of officers who served there in recent times. Some served as Adjutants or Captains and came back as Lt. Colonels to lead the division. Captain Della Rapson was the first officer at 115 Valencia down the street.

The building's architects were well known in the city in 1909 when Charles Paff and John Baur did the design. In the same year they designed the Fugazi Bank at Columbus and Montgomery. Three years later in 1912 Paff also gained recognition for doing the Olympic Club Building at 524 Post.

The Army's building was listed in the State Register of historic buildings in 1994 and is San Francisco Structure of Merit #9. Original construction costs were $78,000.

Original owners were the Knights of Pythias, who moved in during 1911. In those days, America was full of fraternal orders banded together in the spirit of "friendship, charity and benevolence." The Knights called their building the "Pythian Castle." It was their Grand Lodge, a rather grand name considering the basically working class neighborhood in which it was situated.

Before the earthquake of 1906, the area was full of Victorian houses, most of which were destroyed in the fire. Later, when the four-story "castle" was built, it towered over flat frame houses and commercial establishments. In the immediate neighborhood, there

101 Valencia—home of Salvation Army administrative, training and worship services from 1920 until 1989. Photo courtesy of The Salvation Army Museum of the West

were two restaurants, two laundries, a cornice works, auto repair shop, printer, coal and grain shop, wrecker's yard and the machine shop of the United Railroads streetcar company. Lodge members, mostly African Americans, came from throughout the city to meet here.[1]

Although the address is listed as 101 Valencia, the entrance to the building is on McCoppin Street named for Frank McCoppin, Mayor of San Francisco from December 2, 1867, to December 6, 1869.

When the Army bought the property in 1920, it could hardly believe its good fortune. Very little remodeling was needed. The women's dorm was on the third floor. Men were on the fourth, along with the Drill Hall which served as a gymnasium. Classrooms were on third and fourth floor, administrative offices on second.

A Sad Loss on "Silver Hill"

Not anticipating the economic trials of the years to come, in 1928 the Army had big dreams and access to modern methods when they built the training school in the 800 block of Silver Avenue in San Francisco. They saw it as an example to the rest of The Salvation Army world, a training school with all the newest facilities. There was a house on each side, one for the principal and one for the chief secretary.

CHANGING TIMES

National Commander Evangeline Booth came west to dedicate the building that fall just a few weeks after the first class moved in. Young Henry Koerner was in the class. "The ceremony was very, very, very long," he said, laughing with great amusement. "We cadets had to scurry downstairs to clear tables for the evening meal!"

"Cadets, I'm not through yet!" Evangeline admonished the group when she saw them considering whether to leave.

Bare land stretched out on either side of the structure and the steep gulch behind led to the St Mary's section of town. Houses across the street eventually held young children, many of them Italian, who came to Sunday school classes in the building.

The Army had very strict rules for young, unmarried cadets. Their average age was about 18 and many were in the big city away from home for the first time. To observe proper decorum, they put men's rooms at one end of the long building, women's at the other. All meals and all classes were separate, except for assemblies. If they met in the hall, cadets were always to refer to each other by last name. Men were not allowed to talk to a girl in public and had to get permission to correspond, especially if their intentions were marriage. They weren't even allowed to get on the same streetcar going downtown. If a group was on its way to the Citadel at 101 Valencia, women had priority. Men had to take the next car.

During the Depression, the Army suffered the fate of many other property owners. There was no money to keep up payments. In 1933, after six sessions, the building was lost to bond holders.

History of the property is sketchy from that point until 1955 when Simpson Bible College purchased it. Lore has it that the Army Corps of Engineers possibly used it as a detention facility after a fire on Angel Island. During World War II, the story is that a German Countess and her son lived there while he was studying at Stanford. After the war, Christ's Church of the Golden Rule, a spin-off of Mankind United, purchased the building for $250,000. They used it as a seminary until 1954. Then a rancher in Bakersfield sold it to Simpson Bible College.[2]

In 1989, Cornerstone Evangelical Baptist Church and Academy purchased the property and did extensive renovations.

In 2004, as guests of Cornerstone, a small Army delegation was offered a full tour of the buildings, going down and still further down steep stairs to the old kitchen with its ancient dumbwaiter, to the boiler room and to the gym where Citadel and cadet teams used to play basketball. As a child, when her father was assistant principal, Evie Dawe lived near the school in a house on the next street over. It's still standing.

On the day of the visit, almost a hundred Chinese children were practicing for a Christmas pageant. And when they were through, Evie sat down at the piano to give a spirited rendition of "The World for God" by Evangeline Booth. The words and original Salvation Army crest are still inlaid in the marble floor. The hosts pointed out that the floor had been carefully retouched to match the original.

It was a memorable visit, probably the first in many years that Army officials had been in the training school their forbears had built. "How sad they must have been," said Captain Robert Lloyd, "to lose such a fine building."

Brigadier Therma Cline (R) and Lt. Colonel Charles McIntryre (R) say one of their first memories of their training session all those years ago was moving furniture from Silver Avenue back to 101 Valencia. Men cadets were housed at the Mission Street Corps and came by streetcar for daily activities.

The Depression Hits Home

In 1920, the Army bought the building at 101 Valencia. In 1922 the Booth Home and Hospital in Oakland opened. In 1924 the Evangeline Residence for Women opened. By 1926 the Red Shield Youth Program was in place. In 1928 there was a grand new training school on Silver Avenue.

Enjoying new stature because of its work on the battlefront, mortgaging properties and depending on the good faith pledges of its donors, the Army had moved to expand its outreach by selling bonds to finance projects like these.

When the stock market fell apart and the bonds became due, the Army—like the rest of the world—found itself in tenuous financial times. Stories say there were weeks when the Territory was within days of plunging into default. These were brutal times—for everyone.

A story Major Wayne Froderberg remembers from

training school came from Major Minnie Belle Shennan. She was at a civic function in the city when a well dressed businessman accosted her saying he had no respect for the Army. As a boy during the Depression he had eaten in a soup kitchen and had become very ill from the food. He named "the date, place, day of the week and time he ate" and suddenly Major Shennan realized it was her parents who had been the officers responsible for the meal. It was their corps, their town.

She also remembered that her family hadn't drawn a salary in months and that they never sat down until they had fed the last person in line and finished the dishes. They were the last ones to eat and often there was barely enough. She and her sister had eaten the same soup and after years of such fare, she had actually developed a chronic disease as a result.

Salvation Army officers who lived through those years saw the Depression through the Army's own financial difficulties and also through the eyes of every person who ever had to stand in line for a meal.

"Pluck and Go"
What It Takes to Make a Good Officer

It's understandable that early Salvation Army cadets were said to major in subjects like "scrubology, sweepology, bootology, bed-making-ology" and other household work. In the 1890s, their days were filled with study and prayer, yes, but the practical application was work, and lots of it. To raise money for their outreach and quite literally to buy groceries for their meals, they sold *War Cry* periodicals in saloons and door-to-door. Was it raining? Or unseasonably cold? Every day, whatever the weather, they led open-air meetings on the streets.

Turn-of-the-twentieth-century Salvation Army leaders took very seriously the business of getting strong candidates to continue the work. Questions on the officership training application minced no words.

"Do you understand that you may not be allowed to marry until two years after your first Commission as an Officer, and do you engage to abide by this? If you are not courting, do you pledge yourself to do nothing of the kind while you are a Candidate, during Training and for at least twelve months after your appointment as an Officer?

"Have you ever been an open BACKSLIDER? How long? Did you ever use intoxicating drink? How long is it since you entirely gave up its use? Did you ever use tobacco or snuff? What uniform do you wear? How long have you worn it? Can you start the SINGING well and readily?

"Are you short of any front teeth?"

"If so, will you get others put in if accepted?"

Reference forms were just as direct. "Has the Candidate got the PLUCK and GO necessary for an officer?" . . . "Is the Candidate given to giggling or making silly jokes?"

"Don't make the mistake of giving a good reference if you don't really believe it," warned the Army. One unfit person could easily do endless damage.

Candidates were mostly young, some from desperately impoverished situations, some leaving behind a life of depravity and/or demon rum, most looking for a lifetime commitment and a practical way to demonstrate their often newfound faith.

Young women emboldened by women's suffrage found the Army a safe and very satisfying place to exert their independence. As the work continued, children of officers followed in their parents' footsteps.

A Daughter of the Regiment

There were no marketing directors, but early Salvationists certainly knew how to turn a phrase. Everything and everybody had a name—including officers' children who followed in their parents' footsteps. Edith Smeeton Morris was a "daughter of the regiment" (in other words, a preacher's kid).

Her parents were officers who had worked directly with General William Booth in London, then moved on to assignments in Canada and Chicago. In 1906, the general reluctantly posed for a picture including Edith and her brother and sister. As Edith explained to Western Territorial Army historian Frances Dingman in later years, Booth—who was a crusty, dyspeptic old man at the time—worried that if he sat for one picture with children, other families would want one too and when would he have time to work? Added to his concern was Edith's gigantic hair ribbon, which apparently tickled his nose!

Smeeton kept a diary when she was in training at 101 Valencia in 1922. And it's clear during those days that chores were still a primary component of the cadet's workload. Crocheting also seems to have been routine, illustrating that women's work here, as in the rest of the world, was still strictly defined. Work included kitchen duty, dealing with problems of backed-up plumbing in the laundry, ironing, starch and more starch. "Oh, you starch!" she sighed. "Fair and I washed pillow shams and had oodles of fun." "We ironed all day and tonight I am surely tired."

Her *War Cry* report one day showed the girls' average sales were 62½ and the boys' 47½. She sold 50. Later that night, she reported: "Col, Fair, Cox, Harris, Yates and I had oodles of fun in Lieut's room reading 'Catherine Booth.'"

Between classes, washing, ironing, crocheting, doing kitchen duty, leading open-air meetings and selling copies of *War Cry,* the women cadets had a full schedule. There were the stereotyped domestic chores, for sure, but during a time when other churches were far less open, women had opportunities for leadership. Smeeton and others participated fully in public meetings.

Photos in the family's early scrapbooks show the same kind of off-duty horseplay you'd find in any other school dorm of the day. Photographs taken on the building's rooftop show the coeds mugging for the camera.

One picture shows that the group went to Sutro Baths to swim. And on at least one evening, she and a compatriot ventured out to a nearby candy store and listened to the radio. It was "very wonderful," she said.

In Training at 1450 Laguna

During the years at Silver Avenue and 101 Valencia, the few married couples in training had made real sacrifices to join the Army, leaving young children with friends or relatives while they studied.

After the war, more married couples entered Salvation Army training. They came in at an older age, often with more schooling than in the past. And they often had children. A childcare facility was included in the school at 1450 Laguna. Major Shawn Posillico remembers attending Raphael Weill Elementary School (now Rosa Parks) when her parents, David and Effie Patrick, were in training. She remembers the day Martin Luther King was killed and the care that school staff took to get children home safely in the aftermath of violence.

By the 1960s and 1970s, there were new challenges for training cadets. Lt. Colonels Harry Larsen, Victor Newbould and Willard Evans were principals during the flower child revolution.

Opening Day of the Silver Avenue Training School. Photo courtesy of Lt. Colonel Check Yee

TOP: *Training College Parade of Life-Saving Guards, and Guard's Divisional Drum Corps, 1920. Photo from Northern California and Nevada Divisional Archives*
BOTTOM: *Annual Band Festival, Northern and Training College Divisions, 1935. Photo from Northern California and Nevada Divisional Archives*

Like the rest of the world, the Army worked hard to understand changing times. To speak to issues of racial prejudice and learn from each other, Colonel Newbould made sure cadets visited a wide variety of churches, synagogues and Grace Cathedral. Captain Ray Dexter led sensitivity training. Cadet Wayne Froderberg remembers taking "Quotations from Chairman Jesus" with him as he went out to talk to people on the streets. Cadets Joe Noland, Doris Tobin, Marilyn and Ron Bawden called themselves Salvation Singers and performed in North Beach cafes, along with Major Sharon Robertson. And everyone who was there at the time remembers Dot Larsen baking delicious hot cross buns in the kitchen.

Ultimately, the impact of cultural changes on cadets looking to fulfill their vocation took its toll. Class enrollment dwindled. The 1971 session with Cadet Joe Posillico had only seven single cadets and five couples. In 1974 Cadet Tom Rainwater was killed by the infamous Zodiac Killer while standing on a street corner. Before Redevelopment improvements, drug busts and unrest plagued the neighborhood. It became harder and harder to attract cadets with families.

It was a tumultuous time. There were innovations locally, but also thoughts of moving to other areas, including the possibility of cooperating with the University of the Pacific in Stockton to share libraries and social service training. When an extremely suitable piece of property became available in Southern California, the Army relocated its training school.

Masasuke Kobayashi, Soul of the Japanese Division

Major Masasuke Kobayashi was a gifted speaker, dynamic leader and very effective conduit of Salvation Army services to the Japanese community in San Francisco and other parts of California. He was the prime mover in cultivating The Salvation Army's Japanese Division in San Francisco and was highly respected for his work.

As a Presbyterian minister from Salinas, California, Kobayashi heard a Salvationist speak at the San Francisco World's Fair in 1915 and was so moved that he invited the Japanese Commissioner Gunpei Yamamuro to speak to immigrant communities on the coast. Kobayashi became a Salvationist, traveled to Yokohama for officers' training and returned to the west coast to establish the Japanese Division of The Salvation Army.

Major Masasuke Kobayashi

Work began in the City in July 1919, when Kobayashi and five others reportedly visited virtually every Japanese home in *Nihonmachi* (Japan Town). The campaign was a brilliant success. A month later, on August 16, five hundred Japanese attended the official opening of the Army's Japanese Division in the United States. Within the next year and a half, twelve officers, 18 non-commissioned officers and 208 soldiers enrolled. Eventually Japanese corps opened in Sacramento, Visalia, Seattle, Oakland, Stockton, Los Angeles, Fresno and San Jose, along with a Home of Rest, medical clinic, children's home and day-care center in San Francisco.

The Japanese Divisional Headquarters was dedicated in 1921. And on February 28, 1937, after a five-year fund raising campaign, the Army opened a new building at 1450 Laguna Street. This Salvation Army program was supported by the Japanese business community, which raised about $61,000 for the building. The Emperor of Japan donated five thousand yen—in 1931 dollars, about $1,750). The Japanese Division with its eight corps was an active part of the Salvation Army.

Kobayashi was such a powerful leader that when he suddenly died of a heart attack at age 57 on October 15, 1940, there was no senior officer immediately ready to take his place and the Division had to be closed. His funeral at The Salvation Army's Citadel auditorium at 101 Valencia was filled to the rafters with people who had admired his work. Colonel Henry Koerner, who handled financial matters for the Japanese Division as well as the Territory, remembered the lobby was filled with flowers. In fact, someone counted them. There were 451 floral tributes.

Sadly, a year later, the world changed dramatically for Japanese-Americans.

Although they made a valiant attempt, there was little The Salvation Army could do to save its Japanese Division from Executive Order 9066 which sent Japanese to the horse stalls of Tanforan and ultimately to Heart Mountain, Manzanar and other relocation camps. Although Japanese Salvationists continued services in the camps, they scattered after the war and never recaptured the Kobayashi spirit of the early years.[3]

Japanese who served as Salvation Army officers until retirement were Major and Mrs. Tozo Abe, Mrs. Brigadier Masahide Imai, Major and Mrs. Ainosuke Ichida, Major and Mrs. Takamaru Hirahara and Major and Mrs. Naoji Suzuke.

Later, the Army used the vacated building as a training school for cadets. After 1954, additional land was purchased and other buildings were constructed. When it was put up for sale in 1975, *Nihonmachi* residents protested, reminding the Army of Kobayashi's strong role in construction of the original building. Territorial Headquarters paid tribute to his memory, and later made a major contribution to the Japanese Cultural Center in Los Angeles in his name, but by this time the Army had no resources or personnel to continue a separate Japanese Division. Finally, the building was leased to "His Way," a small evangelical organization whose dreams apparently were bigger than their pocketbook. When the group was a year behind in its payments, the Army sold the building to John Foggy, a developer who then sold it to the People's Republic of China.[5]

The Salvation Army Tried To Protect Japanese Orphans, But Appeals Were Denied

When the evacuation of Japanese was ordered after Pearl Harbor, The Salvation Army refused at first, hoping to save its Japanese orphans from the internment camps. Requests were denied. The Army tried another solution—moving them from the orphanage at 1450 Laguna in San Francisco to Lytton Springs in Marin County.

Approximately 22 children moved to Lytton on December 26, 1941. Within the first few days, some left to go to relatives. Then Lytton was declared to be in the Priority I area and the Army was ordered to transfer the children again. The superintendent evidently did all he could to keep them, but many appeals were denied. Some children left January 12th. The last group left March 21, 1942.

In a 1973 memo to Colonel Lawrence Smith at Territorial Headquarters, Lt. Colonel Margaret Cox said it had been her responsibility to see to an orderly transfer. She remembered that they tried to find relatives living outside California and in some cases were successful. Other children had to be returned to the Public Welfare Department responsible for them.

"We had no alternative," she said, "but were given a deadline, which had to be met. It was a most difficult time, especially for the children." Many of them had lived at The Salvation Army's Japanese Children's Home for years.[23]

Coming Home Again

Forty-two years later—after the war, after internment, after resettling into American life—people who once were residents of The Salvation Army's Japanese Children's Home returned to San Francisco in 1984 to remember the positive times of their growing up.

Representatives from The Salvation Army's Territorial and Divisional Headquarters worked closely with a group of Japanese-Americans to reconstruct a reunion of people whose lives had drifted far apart. Fifty people attended, including spouses and long lost friends. It was a joyous, spirited, poignant reunion.

Overarching Principles

So where does The Salvation Army fit into the larger San Francisco picture? How has such an admittedly conservative organization become so integral a part of such a left coast town? One person remembers how often The Salvation Army has been on the cutting edge of social services. Another says, "Well, more like slightly behind the curve."

Who's right? It depends on who you ask, and when.

Former National Commander Robert Watson brings it all down to this:

"What ultimately gives you the hope of influence is not that the world is predictable and controllable and that, therefore, all you need is the right strategy to control it—because we know that's not true. Rather, the hope is in adapting to constantly changing events in ways that keep you aligned with unchanging, overarching principles."[6]

From "Unwed Mothers" and Adoptions to Teen Parenting

The Army has been in the business of providing sanctuary for pregnant women from its earliest years. Long before there was the International Soundex Reunion, before there were organized support groups for the adoption triangle of adoptees, birth parents and adoptive parents, the Army was there.

Booth Home in Oakland opened in 1922. In the beginning, it was a rescue home for destitute women from the streets, many of them pregnant. Later, it was a safe haven for young unmarried women hiding their pregnancies.

"Dear Abby" used to run letters from women saying how grateful they had been for the Army's help during a difficult time. Their impression of a home for "unwed mothers" changed when they arrived. It wasn't a place where "bad girls"—some as young as 12 or 13—went to have their babies. It was a safe place for pregnant young women with no other place to turn.

Later still, it became a residential program interacting with the school system, reaching out to younger and younger teenagers who kept their babies.

Society was changing. And the Army was, too. The program's emphasis now focused on life skills and choices. "Take Charge of Your Life," a workbook written by staff in the 1980s, stressed the importance to teens of making wise decisions about their own lives before accepting the responsibilities of parenthood.

For years, the Army kept its promise of anonymity to women who relinquished babies during the middle part of the century. Today, as adoption laws have changed, the Army is also adjusting. Client records for maternity homes and hospitals have been centralized and non-identifying information is available to both birth mothers and adoptees so long as both give permission to release it.

The Army plans to follow up with counseling, guidance and referral for at least a year after a reunion of birth mother and child. All research on the West Coast is done through The Salvation Army Missing Persons' Service—11th Floor, PO Box 22646, Long Beach, CA 90801-5646. Call: toll free (800) 698-7728 or direct (562) 491-8321.

Adela Rogers St. John Pretended to Be Unemployed

At the height of the Depression, so the story goes, William Randolph Hearst asked star reporter Adela Rogers St. John to do a series on unemployed women. She dressed the part and hit the streets. Ultimately it was a clean bed and clean sheets of a Salvation Army low-income residence that she remembered most gratefully.

"I couldn't be more grateful for the chance to say how greatly I admired Evangeline Booth and The Salvation Army," she said. She picked the Army as the only charity organization she could recommend 100 percent.

In her autobiography, The *Honeycomb*, she spoke of the warmth of Salvation Army hospitality. "I don't know how they could love me, but they did. The lassies had breakfast with me and cheered me on my way."

The Evangeline—A Home Away from Home

You don't have to be cut in the mold of Jack London or Jack Kerouac to know what it's like to be on the road, to be in a strange city and on your own.

In the old days, when sweet young girls migrated from small towns to the wicked city to find their fortune, The Salvation Army built low-rent Evangeline Residences for young businesswomen. They were clean, safe, strictly supervised and for women only. San Francisco's Evangeline opened in 1924, an eight-story, 200-room women's residence at 44 McAllister.

Majors William and Edith Morris supervised the program. When he died, she was commissioned a brigadier in her own right and continued in her job until the program was ultimately closed.

Changing times had taken their toll. Hippies moved to the Haight and communal living situations. Liberated businesswomen looked for upscale independent housing. By 1981 it had became increasingly "uncool" to live in residence hotels, even though the cost was as little as $250 a month. The neighborhood itself was changing. For all the safety precautions the Army built into its residence just off Market Street, there was still the gauntlet of loiterers on the street to be faced. In spite of the amenities (being close to work and downtown shopping, proximity to city-wide transportation, getting free tickets in lieu of pay for ushering in local theaters) there was always a permanent band of street people who could easily follow residents' movements.

Thirty years later, a woman tells of being sexually assaulted by a man who had followed her on Market Street. Terrified, she finally got back to the safety of the Evangeline, went to the manager's office before daylight and, with her help, got the first plane out of town. "I was so surprised to see the officer up so early," she remembers years later, still obviously shaken by the incident which appears to have deeply affected her life. "The Salvation Army was wonderful."[7]

Eventually an older population moved into the Evangeline and they, too, were vulnerable. For a few years, vacancies at the residence ran sometimes as high as 50 percent. Eventually the number of unused beds and increased operating costs necessary for major repairs to meet building codes convinced the Army to sell the building and channel monies into new areas.

The Evangeline tried hard to give residents a sense of family and community with group activities and celebrations. Edith and William Morris were married there. Years later their son Art was married on the same spot. Art reports that Liberace was a guest at the second wedding.[8]

The Haight Street Years

The period between 1961-1976 were years of dramatic changes in the world—flower children came to San Francisco, drug use increased, grass roots agencies emerged. The Civil Rights Movement made bold strides. United Crusade became United Way of the Bay Area and adopted the slogan, "New Directions." The Salvation Army's Northern California and Nevada Divisional Headquarters moved from Oakland back to San Francisco.

These were the Haight Street years.

Territorial Headquarters still held the fort at 101 Valencia, where it had always been. And now there was a well-worn path between the big red brick landmark and the 60 Haight Street Divisional Headquarters just beyond the freeway on the other side of Market.

Lt. Colonel Ranson Gifford, Lt. Colonel William McHarg, Lt. Colonel Richard Angel and Lt. Colonel Victor Newbould were divisional commanders.

Because of its network of services in so many communities throughout the United States and over one hundred countries, the Army often receives notices of missing children or relatives who may have ended up on the streets. In the hippie era, the media sometimes referred to them as "throwaway teens" . . . children whose parents gave up trying to understand the Sixties Generation.

These young people were the heartbreakers, reported Colonel Pauline Eberhart who was in charge of the Army's Western Territorial Missing Persons Department during those years. She worked closely with Huckleberry House, The Switchboard, Off-Ramp and other youth groups. She was famous for taking off her bonnet and rapping with teens.

Carol Doda

The image of topless dancers Carol Doda and Yvonne D'Anger wasn't just the larger-than-life neon sign outside the Condor nightclub in the 1960s, it was thoroughly implanted in the popular culture.

When a small band of Salvationists marched through the streets to take a stand against the activities in North Beach, some onlookers were curious, others disdainful.

In a way, it was like the Army's earliest days in San Francisco crusading against the wickedness of the Barbary Coast. The wire services loved it.

A crowd of several thousand heckled the Salvationists, said United Press International. "Take it off," they shouted to Salvation Army lassies in their tunics, which in those days had stiff stand-up collars buttoned tightly at the neck.

Associated Press put it into historical context. Fighting sin and degradation in the days of the Barbary Coast was one thing. But in contemporary society, San Francisco actually encouraged pride in its bawdy past. Images of lusty dance hall girls and rowdy gold miners were part of the tourist package![9]

Devil's Island

Prison work has always been part of the Army's outreach. Commiaaioner Charles Paen's book, *The Conquest of Devil's Island*, describes in detail the penal colony in French Guiana, where shark-infested waters, swamps, disease and an unjust sentencing system in effect doomed all convicts—no matter what their crime—to life imprisonment.

For 20 years, the French Salvation Army was relentless in calling for reforms, providing jobs, housing and food for prisoners who were forgotten and left to die in that desolate area. The Army was a major force in causing the settlement to be closed and in 1953 when it finally happened, the Army took the lead in prisoner repatriation.

The Brighter Day League

Working with prisoners was a major effort in the United States as well. In 1930 the San Francisco Salvation Army had the largest "Brighter Day League" in the country. A total of 1,905 prisoners in San Quentin were enrolled.

During the 1960s in local courts and jails, Envoy Agnes Nightingale was a highly-regarded ombudsman for young women on the other side of the law. Before sentencing, a judge would turn to the lady in blue and say, "Call her parents. Tell them what's happening."

After her came Captain Jeanne Rigney who responds with an enormous roar of laughter to remember a judge looking at her and a young woman released into her custody. "If you run," warned the judge, "you're *both* going to jail."

After they left, the girl asked if she could use the restroom at a filling station along the way. Rigney agreed and, you guessed it, the girl slipped out the window and was gone in a flash. Rigney spent a sleepless night wondering how to face the judge. She tells the story with such ingenuousness you can imagine she arrived in front of the bench with her hands outstretched, ready for the handcuffs!

Years later the girl found herself back in jail, in front of the same judge and the same captain . . . "Aha," says Rigney. "She knew better that time!"[10]

Next in the Army's northern California prison ministry came Major Buzz Brewer who because of his height and demeanor was dubbed "the gentle giant." He flew his own plane to prisons throughout the Division. Envoy Dorothy Steele, Majors Gerald and Suzanne Hill, and Envoy Elsie Lantz followed, each putting a personal stamp on the job. Lantz started a Home League in the County Jail and with one very special inmate started a drive to crochet clothing and blankets for preemies in the HIV ward at San Francisco General Hospital. "Those babies need a hug!" declared the lady in jail. She knew what it was like to be away from your children.

In 1996, KPIX-TV collected over 100,000 books for The Salvation Army to distribute. Officers reported that thousands of them went to local jails, prisons and rehabilitation programs.

Saturday Night Coffee House

During the 1970s, Major Sal Gomez and Cadet Moses Reyes from San Francisco's Corps/Community Center in the Mission welcomed up to 75 and 80 neighborhood teens at their Saturday night Coffee House. Special attractions included groups like "The Saints" rock group from the Modesto Corps and ex-Black Panther James Weston, who had spoken on several campuses around California on personal identity and race consciousness. The Corps' Wednesday afternoon "Fun Factory" attracted an average of 40 to 50 children looking for organized after-school activities.

You Can Be Groovy

A 1968 training school recruiting poster hung in several Salvation Army coffee houses which sprang up across the country, featuring live rock bands and reflecting the times. Appealing to young people looking for a vocation, but still trying to be hip, the poster encouraged new recruits by saying: "You Can Be Groovy and in Gear and Gassy and Outta-site and Unhung Up and Wild and Up Tight Ringer and Tuned In and Cool and Turned On in a Very Special Way."

Notes

[1] Report prepared by William Kostura, architectural historian, for Corbett and Minor, with contributions by Michael Corbett and Zachary Nathan, 1995.

[2] Excerpted from "An Extremely Short History of the Building That Simpson Occupied in San Francisco" by Miles S. Compton, Simpson University Archivist.

[3] References: Frances Dingman, Salvation Army Museum of the West, phone interview with Mrs. Major Florence Abe in 1997; Henry Koerner; Renee Tawa, *Los Angeles Times* (March 11, 1997)

[4] References: Frances Dingman, The Japanese Division, Salvation Army Museum of the West; *Hokubei Mainichi* (Nov. 15, 1975); *SF Examiner*, (Jan 26, 1976); *SF Chronicle* (Jan 27, 1976); *SF Examiner-Chronicle* (Aug 13, 1978); *Hokubei Mainichi* (Nov. 17, 1979)

[5] 1973 memo from Lt. Colonel Margaret Cox to Lt. Colonel Lawrence Smith provided by Frances Dingman, Salvation Army Museum of the West

[6] "The Most Effective Organization in the U.S." by Robert A Watson and Ben Brown. published by The Salvation Army National Corporation, 2001. pp 176-7.

[7] Interview with former resident, 2003

[8] Material drawn from interview with Art Morris, 2003, and *San Francisco Examiner* articles

[9] United Press International and Associated Press reports, September 25, 1965

[10] Interview with Major Jean Rigney at THQ, 2003

FIVE

The Act of Giving

Tzedakah Box

In Hebrew, the act of charity is called *tzedakah*, which translates as "fairness" or "justice." The Jewish commandment, *tikkun olam* is an obligation, calling for people to share with God the work of "repairing the world."

So a *tzedakah* box—whether in the First Temple in Jerusalem, a *pushke* in the *shtetls* and ghettos of Central and Eastern Europe before World War II, or the familiar Jewish National Fund blue tin box in nearly every Jewish home in the 1950s—is a container with a slot on top to collect coins regularly and anonymously for the poor, a reminder of our responsibility to others.

A container with a slot on top to collect coins for the poor. That has a familiar ring . . .

It sounds like the great granddaddy of the alms box in the cathedral. In spirit, if not a direct descendant, it sounds like the cardboard missionary box given to Sunday school kids at the Swayzee Methodist Church in Indiana, like a collection box for The Salvation Army's annual Self-Denial campaign, a March of Dimes can or a UNICEF box.

Carrying the concept to what would seem a logical conclusion—it sounds a lot like a Salvation Army red kettle!

In an era when people itemize their charitable donations, when mail appeals proliferate, when megafundraisers often gross a lot and net remarkably little, when the national debt and Federal budgets are in the billions, the Army's stalwart red kettle still stands on the street corner collecting coins.

Usually a low-income person being paid a minimum wage stands with the kettle. It's not an easy job and in the larger scheme of things, it doesn't raise a lot of money. In a world of mail appeal, the kettle represents "a drop in the bucket," a gentle reminder that you might want to go home and write a check or volunteer to help. There's still much more to do.

In a time of year when consumers spend an extraordinary amount of money on holiday gifts, the tiny silver bell and red kettle remind them—much like the *tikkun olam* commandment—to help "repair the world."

Pass It On

"Bestow charity in such a way that the benefactor may not know the relieved persons, nor they the names of their benefactors . . ." That's the way Moses Maimonides saw it in the 12th Century. Major Ron Toy owns a Salvation Army good deed token from 1938 with the same message. It's a small metal token with the inscription, "That your charity may be secret and your Father who sees will reward you." It appears to be a kind of receipt for assistance given and a reminder to pass it on. The message reads:

> Don't thank me!
> I have paid a debt
> For help received.
> The obligation is now yours.
> *PASS IT ON*

Major Donor Known for His Excessive Thrift

"OF HUMAN INTEREST: Around the Sir Francis Drake Hotel on Powell, the staff used to chuckle about Kasson Avery. 'What a character!' they'd say in the most loving kind of way.

"A bachelor, he was one the Drake's two permanent guests, having checked into Room 803 ($300 a month) shortly after the hotel opened in 1928. There was nothing unusual about his appearance: tall, slender, neatly dressed. And he was exceedingly polite.

"What made him a 'character' was his excessive thrift. The staff was amused that every night, he'd walk up and down the halls with a paper bag, picking up empty bottles that the other guests left outside their doors, and taking them to a nearby grocery store for cash refunds. And he could never pass the phone booths in the lobby without checking the return slots for coins.

"Now and then he'd approach David Plant, the Drake's general manager, and wonder nervously if his rent were about to be raised, 'what with inflation and all.' 'Don't worry about it, Mr. Avery,' Plant would reply kindly. 'We are pleased to have you as a permanent guest.'

"Well, the careful Mr. Avery died at the age of 87— and left a net estate of $1,848,000. Under the terms of his will, The Salvation Army gets $924,000 and the YMCA and YWCA $462,000 apiece—much to the startled delight of the officers of these organizations. They had never heard of Kasson Avery.

"Footnote: After his death, the Drake's housekeeper was cleaning out the drawers in his room and found a little black book in which he had entered his living expense for the past few years. Never once had he spent more than 35 cents for breakfast or $2 for lunch or dinner."

—*Herb Caen, SF Chronicle, July 21, 1970*

This Little Light of Mine

It started small—one man's offer to help children infected with HIV/AIDS—and, just like the song, Lyle Richardson's "This Little Light of Mine" project has continued to shine. The 81-year-old baritone learned about The Salvation Army's role in helping HIV-infected babies in 1999 through Major Isa McDougald. The babies had been abandoned in a government hospital in St. Petersburg, Russia. Since then, he's raised thousands of dollars to upgrade bathroom and laundry facilities in the hospital and, most important, to make sure Salvation Army workers are there to give hugs and emotional support to the children.

The project has become a personal crusade. He leads tour groups to the hospital and, with great panache, sponsors elegant fundraising teas and music recitals in his home. In St. Petersburg, he visits The Salvation Army and sits down at the piano immediately to lead the singing with the local Home League women's group.

Their language is Russian. His is English. It makes no difference. Music transcends the barrier. The room is filled with harmony.

On the tour boat—at the grocery store, at the post office, at social events, in the retirement village where he leads the choir—whenever and wherever he can, he talks passionately about "his babies."

With the help of professional colleagues, Presbyterian churchwomen, friends, cousins and even the encouragement of Sarah, Duchess of York, he's proof that one person's commitment really can make a difference.

One Donation at a Time

One person, in the quest for frequent flyer miles, buys 12,000 cups of pudding and donates them all to charity. One corporation decides not to light its building at Christmas and gives the money to help buy food for poor families instead.

One organization like the United Anglers Enforcement Committee turns over a day's fishing catch to the Army once a year. Larry Silva and fishing friends leave port before sunrise and return at the end of day with a boatload of fish, not for themselves but for addicts in recovery.

A woman gets out of a sick bed to come downtown late Christmas Eve when other office workers have already gone home. Sneezing and wrapped in many layers of clothing, she waits in the lobby of her office building for Army volunteers to pick up a ballerina tutu she has made for a young girl K101 deejay Don Bleu has talked about on the radio.

Another buys a special doll and says she knows the Army will give it to a girl who will really appreciate it. The OMI Business League finds that child at a small neighborhood community center behind a barbershop on Broad Street and the look on her face is incredulous...!

One donation at a time. That's what it's all about.

The Governor's Wife Baked Pies

In the 1930s, Salvation Army Home League bazaars filled the Citadel auditorium with needlework and apple pies. And eventually they became elegant social events. By the 1950s and 1960s, they moved out of the 101 Valencia Street auditorium into the Native Sons and Larkin Hall Auditoriums. Officers like Pauline Eberhart, Elnora McIntyre, Wilma Smith and Edith Morris were deeply involved and distributed promotional flyers widely to nearby government offices. Often they raised $8,000 in a single day.

For several years, Mrs. Earl Warren, wife of the Governor, brought cakes and pies to sell. These were so popular, even the pie plates were auctioned off. Society matrons brought their personal tea services. Guests, of course, wore hats and gloves.

Raising Money With Gentle Hands

Forget the hats. Forget the gloves. Today's bazaars, like most public events, are much less formal.

In the 1970s, the San Francisco Women's Auxiliary was organized through the efforts of Lt. Colonel Ardys Newbould and Major Lucille Youngquist. Joan Morris was first president and, as wife of the Advisory Board chair Henry Morris, enlisted the participation of other Board wives and friends for five years until they were granted their first charter.

Initial members included Sophie Hoffman, Dorothy Kislingbury, Mary Jane Lee, Barbara Rolph, and Evelyn Mulpeters. Then Barbara Ann Lee joined and she brought a strong contingent from San Francisco's Chinatown. Together these women have volunteered for 20 to 30 years as the core group, along with active new members who since have joined. They're busy grandmas and great grandmas now. Some walk a mile or two every morning for exercise. Some take memoir writing classes. But with remarkable loyalty they still meet every Tuesday to plan and implement two annual fundraising events—the Christmas bazaar and the Easter Omelet Luncheon.

Average annual results—$20,000.

The money goes a long way. They've provided groceries for holiday meals for inmates of local prisons and sent Christmas gifts "from Mom" to their children. They've taken children from low-income families on holiday shopping sprees as well as provided sewing machines, pianos and supplies for city units, camperships and even a petting zoo for Camp Redwood Glen.

Members produce gifts and decorations with materials other people cast aside. That's their secret. They're the ultimate recyclers. Pressed flower bookmarks and greeting cards, Christmas trees decorated with 1950s clip-on earrings, elegant ribbon-covered shopping bags, crocheted pot scrubbers, stuffed dolls, hand-knit baby sweaters and elegant angels . . . By hand and with enormous care, they deftly craft the most unusual Christmas gifts their imaginations can design. And thirty years later, Joan Morris is still making prodigious amounts of nut bread and chutney.

SIX

Christmas and the Popular Image

Who's Making All That Noise?
Volunteers Make Celebrity Bell Ringing Street Theater at its Finest

Celebrity Bell Ringing Day in San Francisco's Union Square—
Sid Colberg and Ann Moller Caen with the Herb Caen Memorial All Weather All Brass Band.
Photo by Russ Curtis

On the day after Thanksgiving in 1954, hotelier Ben Swig kicked off a Tree of Lights Fund for the Army with a rally in San Francisco's Union Square. It was important to give thanks "not just on one day, but every day," he said. And every time someone made a donation to The Salvation Army, a bulb would be lit on the tree. He wanted to have so many lights it would "light up all of downtown."[4]

That was a rich prelude to the days his future daughter-in-law and the City's premiere fundraiser and her friends would arrive. San Francisco's favorite mover and shaker Charlotte Mailliard then Swig then Shultz conceived and nurtured Celebrity Bell Ringing Day in Union Square. In the tradition of Amelia Kunkle Devine, she saw the potential of a tiny bell. She and her "A list" band of community leaders kicked it off in 1980. Legendary *San Francisco Chronicle* columnist Herb Caen reported that the "indefatigable and wholly remarkable" Charlotte and her friends collected $6,300 in their Salvation Army pot showing that Army bells signaled the beginning of the Christmas spirit. "People, he said, "are so accustomed to hearing horror stories that we forget the simple fact that most people are okay, given half a chance. Christmas gives them that chance."[5]

Charlotte's own special touch kept her bell ringers coming back every year.

Was it raining? She never flinched.

When her right arm was in a sling, she rang with her left.

Advisory Board member Fifi Holbrook *wih event founder and Honorary Chair Charlotte Shultz.*
Celebrity Bell Ringing Day volunteers wear red to make a unified appearance ringing bells around Union Square.
Photo by Russ Curtis.

90   THE BELLS OF SAN FRANCISCO

KGO talk show host Ray Taliaferro and Congresswoman Nancy Pelosi. Photo by Russ Curtis.

Was a friend getting married? She played the wedding march as they arrived. Was she on her way to Japan? She took invitations with her to sign personally during the flight.

There was always a personal touch. Major Grace Phillips once was up beyond midnight the night before the event, hand crafting name tags with holly and ribbons for each of Charlotte's friends.

As her responsibilities as San Francisco's and later California's Chief of Protocol grew, Charlotte ultimately assumed the title of Honorary Chair. In 1987 she asked veteran TV personality Terry Lowry to take her place as Chair.

It was an easy choice. Terry and her husband Fred LaCosse had rung bells from the very first bell ringing day in 1980, when they hosted "AM San Francisco" on KGO-TV, (now ABC 7).

With Terry and Fred's increased involvement came more emphasis on media participation. The weather was right. The mood was great. There were Celebrity Bell Ringing Days in the 1990s that had celebrity delegations and remote broadcasts from almost every local station. Ginny Yamate, former Community Relations Manager at KGO-TV, ABC 7, described it best. "People from every station could all get together without competition."

The first event in 1980 was held at the Hyatt on Union Square. After that I Magnin put out a canopy and piano where Charlotte's "celebs" like songwriter Sammy Cahn sang with Noah Griffin. Dan Sorkin, Mal Sharpe, Mike Pritchard, Karl Sonkin, Bernie Ward, Harry Osibin and Ray Taliaferro shilled for Charlotte's friends, media personalities, congresswomen and mayors. Ann Fraser and Ross McGowan from KPIX were a popular duo in those early days. Also Frank Dill and Mike Cleary from KNBR. In the days when deejays spun platters of popular music and encouraged listener feedback, Gene Nelson and Carter B. Smith were a terrific draw.

In front of Macy's you could find a twenty-five foot Togo sandwich to feed the crew. Sid Colberg, Bev Feriera, Ed Archer, Police Chief Earl Sanders, Frank Alioto and

CHRISTMAS AND THE POPULAR IMAGE

"Yeow! What a holiday happening!" wrote a contemporary music station after their whole team had spent the day on the corner. They enjoyed themselves beyond expectations, sang carols in public and went back to the airwaves singing.

Checks often came to the kettle for sentimental reasons. "This is for my mother," said a radio promotions director. "No matter how little she had, she always gave to The Salvation Army. Every Christmas, I wrote out a check for her because her Christmas was not complete until she sent that in."

"Please make space for me from 10 to 11," wrote a deejay on a local jazz station. "It's very important that I ring a bell tomorrow." My grandmother passed away this last week and The Salvation Army was her favorite charity."

ABOVE: *Celebrity Bell Ringing Chair Terry Lowry.*
Photo by Russ Curtis.

BELOW: *Former Advisory Board Chair Mary Theroux.*
Photo by Russ Curtis.

Royce Vaughn photographed bell ringers like the dapper Mayor Willie Brown in classic cars. Macy's Delia Ehrlich caught the flavor of the event and the spirit of her store by once arriving in outrageously trendy multi-colored stockings. Both the Westin St. Francis and Hyatt on Union Square hosted luncheons. Ed Lawton and Linda Mjellem from the Union Square Association encouraged, even ballyhooed, the Army's event. Planet Hollywood hosted breakfast and baked hundreds of cookies.

The late Laurie Williams (Robin's mom), a stellar bell ringer, chased potential donors into Saks Fifth Avenue in search of generous donations. Mayor Frank Jordan, Al Attles, Dr. Carla Perez, Kay Noyes and Ann Giampaoli egged her on. Spearheaded by Beth Garner, Saks employees took bell ringing shifts throughout the day.

*KRON-TV movie critic **Jan Wahl**, Charlotte in sunglasses and the late Jean McClatchy were a spirited team. Behind them, sporting an old-fashioned bonnet, is Major Evangeline (Moi) Leslie. Photo by Russ Curtis.*

In honor of their grandfather, Harvey Carter, the entire Steve Barkoff family joins Major Ron Toy to ring bells every year at Neiman Marcus. The Salvation Army has watched their children grow up. Like the James Flood family, which brings the newest generation in a stroller, they're instilling a family tradition of giving. Bryan Hemming cherishes that tradition. His family in England knew General William Booth in the early days.

Porter Deese from Stellar Marketing initiated the first *San Francisco Independent* special insert and was a staunch volunteer for years. Jane Morrison and Gimmy Park Li— was there ever such a loyal team? Gimmy picked up her husband-to-be at the airport as he arrived from China, then left him at home while she hurried to Union Square to take her bell ringing shift. The entire KTVU morning team showed up to ring with the Oakland Raiderettes in front of the Big Shiny Red Fire Engine, which also, by the way, chauffeured Javier Valencia's KRON team with Jan Wahl, Jan Yanehiro and Pam Moore. KFRC's Cammy Blackstone invited a fire truck to bring down their donations and on the way home from a run, that's what they did. Dan Noyes from ABC-7, Major Debi Shrum and Matthew Pearce and crew from Family Radio brought their family dogs. More than once, KTVU's Rita Williams took the day off from work to be at this event. Al Hart crooned Christmas carols with the KCBS team starring

CHRISTMAS AND THE POPULAR IMAGE

Marlon Jones, who worked on computers in the Army's IT Department during the day, took his horn with him at lunchtime to play for Celebrity Bell Ringing.
Photo by Russ Curtis.

San Francisco Kiwanis, St. Francis Kiwanis, Mission Kiwanis, Golden Gate Kiwanis, Rotary, Chinatown Lions, Golden Gate Lions, Ocean Avenue Lions, Altrusans, The Exchange Club, Optimists and Soroptimists . . . At the height of service club popularity, as many as 14 clubs would participate. Today club numbers dwindle, but bell ringing stalwarts remain. If there were a Salvation Army Service Club Bell Ringing Hall of Fame, Bob Brorsen and Angelo Taverna and their teams would certainly be in it!

Kettle workers and their bells are often taken for granted. But they've become such a part of the holidays, when they're gone, people notice.

"There used to be a Salvation Army soldier in front of Woolworth's every year," remembers Rebecca Silverberg of the Excelsior Improvement Association. "Woolworth's would put up tiers of poinsettias and the bell ringer would stand in front. That announced the holidays."

The First Quartet Drowned Out Hecklers

Charles William Fry and his teenage sons, Fred, Bert and Ernest, were musicians in Salisbury, England, where James Dowdle (known as the "Hallelujah Fiddler") had been heckled by mobs in 1878.

Fry, a Methodist minister, was much taken with Dowdle's crusading style. He was also an orchestra leader, so he took his three musical sons—plus two cornets, and alto and euphonium—to the preacher's next outing. The intention was clear. First they played to draw attention to the message. Second, they played to drown out hecklers.

It was the first Salvation Army quartet.[6]

What a great idea, thought William Booth. Music got people's attention. Why couldn't every soldier try to learn an instrument? He made it clear he wanted as many officers and soldiers as possible, male or female, to learn to play some suitable instrument.[7]

In the beginning, it was usually loud and sometimes just plain awful. But what a challenge for the troops!

The First West Coast Band
Struggled Through Seven Torturous Lessons

Instrumental music caught on quickly. So when she arrived in San Francisco in 1884, Mrs. Major Alfred Wells was very perturbed. How could there possibly be a Salvation Army without a band? After a chewing out from the missus, her husband got the message, raised $600 to purchase a complete set of instruments from Sherman and Clay and Company and hired an instructor. Twelve soldiers with no musical background whatsoever struggled through seven torturous lessons.

"What horrible faces they made!" Wells remembered sheepishly but with good humor forty years later. "I told them they blew their cheeks out and grew red in the face, and well anyway, after seven lessons and with new uniforms and instruments we went out into San Francisco streets more than ever a terror to the devil and evil doers."[8]

Early Salvationists accepted their musical charges with enthusiasm. By the time the building at 1139 Market Street was opened on August 15, 1891, a *War Cry* journalist wrote the following bit of Victorian prose:

"The band covered itself with honor by its skill in evolving harmonious sounds from their crooked pieces of brass. The bandsmen's breath had to travel through a tortuous, intricate maze of tubes, but finally escaped all right to the outside world, and by the transformation undergone during its travel, created a pleasant sensation on the public ear."

The Next Time You See a Salvation Army Band, Take Off Your Hat"

Salvationists love the sound of brass. And they play trumpets on the streets because that's where their work started. In the beginning, what they lacked in skill was balanced by pure enthusiasm. Today they offer music camps and advanced seminars on writing for brass and

Unflappable, *The Salvation Army band plays on . . . Majors William and Joy Lum, Major Cindy Foley, Major Ron Strickland. Photo by Russ Curtis.*

CHRISTMAS AND THE POPULAR IMAGE

Salvation Army Band near the Eddy Street Shelter.

regularly go on concert tours. The International, National, New York, Chicago, Pasadena and other staff bands play major music halls. In a large corps, as many as twenty or thirty musicians accompany a Sunday morning service.

At a British National Band Contest at the Crystal Palace, George Bernard Shaw said he considered Salvation Army bands "among the best" . . . because of their fervor, Salvationists get more real music out of their instruments than many professional bands."[8]

Much is made of how hard it is to be heard over the din of street noise at Christmas. It's a hard audience. No acoustics. No applause. Sounds just float off into the ozone.

Like holly and tinsel and shopping, however, bands are an integral part of the holidays. Along with kettles and bells, thrift store trucks and alcoholic rehab programs, bands are the Army image many know best.

In San Francisco, at Christmas, a crack-of-dawn crew greets early morning commuters at the Ferry Building. After their day jobs, they turn up for band practice in the headquarters dining room. Later, they play for first nighters in front of the Geary Theater or Opera House. Has the accountant finished his accounts? The social worker her intake? The IT technician his work? There's music coming from the dining room at Divisional Headquarters. Is there another workplace in America with such a built-in Christmas carol sound track?

For many Salvationists, being part of a street corner band is an understood part of the assignment, no matter how far up the career ladder they go. If a Salvationist can put her lips to a euphonium and blow, she often takes a turn at early morning band duty.

You don't know what a euphonium is? Ask the next Salvation Army musician you see.

Former Advisory Board Chair Sophie Hoffman with Major Chris Buchanan, KRON Public Service Manager Javier Valencia, Lt. Colonel Bruce Harvey and high school students who always helped with the Food For Friends Drive that ran for nine years.

CHRISTMAS AND THE POPULAR IMAGE

At Christmas it's filled with toys. Nancy Udy, Joye Storey, the Easter Bunny (Mary Bishop) and Claire Dunmore at the Jessie Street Service Center. Photo by Russ Curtis.

John Phillip Sousa understood the passion with which Salvationists blow those horns. With a gesture that any well-dressed gentleman in his day would know as a sign of respect, he reminded people, "The next time you hear a Salvation Army band, take off your hat."[9]

The Tournament of Roses Parade And Other Intrepid Horn Blowers

From the six-mile Tournament of Roses parade—where the smell is an overwhelming bouquet of jasmine, onions, lavender, bougainvillea, hydrangea and an occasional whiff of horse manure—to San Francisco's Tenderloin where the stench of urine often permeates, The Salvation Army is no stranger to the street corner.

The New Year's Day parade is the Army's best known public appearance. And it's pure theater. Salvation Army bands love to perform. Musicians travel from many parts of the country to participate.

There's a pride of belonging, a feeling veteran bandsman Dr. Robert Docter, O.F. (Order of the Founder) describes as the same every year, no matter how many times you've played before: "Coming up through a canyon of people who've slept out all night in order to get a front row seat. . . swinging out onto Orange Grove with the cadence. . .starting to play and be a part it all. . ... it's magnificent, a fantastic way to witness."

Remembering Rotten Tomatoes of the Early Years

From the very beginning, Salvationists took their message on the road.

To understand what brought them to the streets, it's important to go back to 19th-century England. William Booth and his Salvation Army in outrageous military surplus uniforms caused no end of confrontation on the streets of London.

Remember, the squalid housing available to the poor in the East End was wretched in the extreme. The corner pub was just about the only place a bloke could go to get away from it all. The liquor industry flourished. People with a taste for the suds drank themselves into oblivion and the next night came back for more. They were steady customers.

Then along came William Booth preaching the evils of demon rum, deploring its effect on battered wives and abused children. When people heard him preach and fell to their knees vowing to drink no more, saloonkeepers lost good customers.

Unhappy about the do-gooders on the street corner, the industry hired "Skeleton Armies" to hound Booth's troops. These were rowdies and street toughs. Sometimes they were hired by tavern owners. Sometimes they were just idlers and hecklers along for the ride, ready to ridicule the Salvationists. They threw garbage and epithets and other rotten things at the evangelists.

In America it was the same. Most people simply didn't know what to make of this strange new Army.

It Was Against the Law To Play a Bugle after Twilight

What people did know was that the Army reached out to unsavory characters that were a nuisance on city streets. City fathers decried the noise and declared it against the law to hold a parade, to play a bugle or to preach after twilight. Laws differed from town to town, but the message was clear. Salvationists were not welcome to declare their faith on the streets.

Think of soldiers, usually sweet young lassies, marching alone through the streets at the turn of the century. Eggs and tomatoes were enemy ammunition. The girls considered it a badge of courage to face the crowds. When physically attacked, they had to get back to the garrison the best way they could.

Pass the Potatoes, Please!

Street corner evangelism was never easy. In Napa, California, however, old timers told a story with a particularly satisfying twist. One night, officers had only 15 cents and no other money for supper.

Like medieval mendicants, Salvationists were responsible for their own keep. There was precious little or nothing in the way of salary. No money in the collection meant no food on the table. So before going out on the street, they occasionally added to their prayers a practical plea for something to eat.

This night seemed different. Was their sermon more rousing than usual? Were the crowds rowdier than the

A street carnival on Turk Street. Photo by Judy Vaughn.

night before? Amazingly, malcontents threw potatoes and onions. At the end of the evening, the thrifty Salvation Army officers simply picked them up, bought 10 cents worth of meat (at 15 cents a pound) and a 5-cent loaf of bread. Their meal that night—a feast of steak and potatoes smothered in onions.[10]

March of Witness

The Salvation Army's annual September March of Witness through San Francisco city streets is a vivid 21st century continuation of a tradition that started years ago.

Year after year, as many as 200 musicians join local soldiers in a march that winds its way through Chinatown and usually ends with a rally in St. Mary's Square. In early years, the rallies were literally held in the middle of the street.

Tourists and members of the Chinese Christian community participate. The Police Department notes the parade organizers' attention to detail. Lt. Colonel Check Yee(R) remembers that police officers told him they liked the parade because it was so positive and not "anti-anything."

In San Francisco, there are a number of groups who take their witness to the street. The variety is symbolic of the diversity for which the city is known.

The Chinese New Year Parade, Martin Luther King March, St. Patrick's Day Parade, *Cinco de Mayo* Parade, Pride Parade, The Salvation Army March of Witness, Columbus Day Parade, Veteran's Day Parade. They all say, "This is who I am. This is what I believe."

Each September, as many as 200 Salvationists from the Bay Area march through the streets of Chinatown.
Photo from the Northern California Divisional Archives.

Cover girl—This illustration for Cosmopolitan *magazine became one of the most familiar Salvation Army images. Many Damon Runyon stories, including the one that inspired* Guys and Dolls, *first appeared in the pages of* Cosmopolitan.

Trumpets at Sunrise

It was Easter morning, April 1, 1934, and trumpeters from the San Francisco Citadel welcomed the sunrise.

This was a special day, the dedication of America's largest cross on top of San Francisco's highest hill (elevation 928 feet). Glenn Gullmes, editor of the West Portal Monthly, has preserved a program showing that The Salvation Army was out in full force.

James G. Decatur, founder of the Easter Sunrise Service, invited Territorial Commander, Lt. Commissioner Benjamin Orames, to help lay the cornerstone. In Washington, DC, President Franklin Delano Roosevelt pressed a golden telegraph key to illuminate the 103-foot concrete and steel structure. In its base were stones from the Garden of Gethsemane, waters from the River Jordan and mementoes from earlier sunrise services held since 1923.

Today the cross, along with the .38 acre area surrounding it, is owned by the Council of Armenian-American Organizations of Northern California. San Franciscans annually climb the steep path to attend services. And Salvation Army trumpets still play.

THE BELLS OF SAN FRANCISCO

The Best of Times
The Worst of Times
The Pictures People Remember Most

The idea that a holiday is supposed to represent family and a time of plenty only emphasizes the fact that for many people it's just the opposite. For some, it's just another day—and not a particularly happy one at that.

Holidays are often the worst day of the year for people alone. Since The Salvation Army redoubles its efforts for those populations during November and December, that means an enormous amount of extra work in addition to programs that continue all through the year. In some ways, holidays at The Salvation Army are totally exuberant, totally exhausting, total chaos.

It's the flat-out best way in the world to understand the spirit of the season.

William Booth learned it first hand. After a particularly satisfying feast within the warmth of his family at Christmas in 1868, he took a walk into the harsh reality of the streets where there were no welcoming hearths and certainly no plum puddings. He made a decision that day that would ultimately become part of the Army's history forevermore. He determined never again to spend a holiday without sharing it with others. In 1869, the Booths served 300 Christmas dinners.

The World's First Christmas Kettle
Was in San Francisco.
Captain Joe McFee Started
A Salvation Army Tradition.

San Francisco, 1891. It was a week before Christmas and the cupboard at The Salvation Army was bare.

Captain Joe McFee dreamed of giving a huge holiday dinner for what he euphemistically called "social derelicts" on the docks in San Francisco. It was a noble idea, without a chance of happening—until he suddenly remembered "Simpson's Pot," an iron pot set out to collect money on a street corner in Liverpool, England.

McFee was an inventive fellow, the kind of public relations scoundrel who would grab hold of a good idea and not let go until he made it work. There was no time to waste. No time for a media campaign. But with the chutzpah of a seasoned professional, he charged into his project and commandeered a cast iron cooking pot. Where best to put it? Why not the Ferry Building at the foot of Market Street near where the Oakland ferry docked? It seemed like a good place to find likely donors—commuters and downtown businessmen.

(Newspaper reporters, take note: This is sometimes reported as having happened in Oakland. Not so. Many ferries docked in San Francisco. The spot where McFee set out the first pot was at the old San Francisco Ferry Building, where the ferry from Oakland arrived.)

On Wednesday, December 16, 1891, *The San Francisco Morning Call* took note of the "novel collection box" and noted that it collected quite a snug sum. The next day, there was another news item: Five "wharf rats," better known for stealing lumps of coal, stole the contents of the pot—$3.60.

Did anybody catch those naughty boys? Probably not. Did the community come to the rescue? Absolutely.

Response to the Army's first community Christmas dinner in San Francisco was overwhelming. Among donations from local markets and friends were a half bullock, seven hams, a sheep, all the fruit, $30 in cash from one firm, all the pork and beans and all the fruit for the plum pudding. The *Call* estimated about 1400 were fed. It was a motley crowd, feasting at three tables, 300 at a time.[1]

McFee's attention-getting, iron pot fund-raising was The Salvation Army's first Christmas kettle, beginning a tradition that spread rapidly throughout the world.

Once a Millionaire

Reporters covered the first dinners in San Francisco after Captain McFee's first efforts. "Keep your eye on this fellow just passing; he is a different type entirely," said *The San Francisco Examiner*. "You can see by the way the shabby clothes are put on, by his walk, and all his movements, that he has been well brought up. Although dissipation has distorted his features and ploughed furrows in his face, the gentleman still peeps through. He was once a millionaire right here in San Francisco." Reporters took note of Salvationists as well, saying, "They are God's messengers and true servants, these humble men and women in bizarre uniform. They dignify themselves and honor their faith by their practical acceptance of the teachings of Jesus."

In 1965, when she was 94 years old, Alice Bourne remembered the first dinner given by the Army in 1891. "We begged every bit of food that was served," she said.

The Tradition Continues: From School Children and The "Swellest" Socio-Athletic Clubs

By 1895, public Salvation Army holiday meals were well established and, according to Army records, well received. In December of that year, a Lieutenant Parke wrote the following review, "San Francisco's Christmas Feast," for *The Conqueror*, an Army publication geared for major donors.

"When San Francisco goes in for anything on the line of public charity, as the South-of-Market Street youngster decoratively expresses it, she 'gets there with both feet.' In no city of all this great and splendid America of ours will you find a finer, readier, broader spirit of largesse; in no city will you meet with a kindlier, jollier responsiveness; in no city will you see all sorts and conditions of men jostling one another in a more amiable emulation."

Captain McFee enlisted children to bring donations to their schools for distribution—groceries, toys, coal, kindling, flour, potatoes and clothing. He gladly promoted a charity football game that brought in $600 in box office receipts from fans of "the swellest social-athletic club of San Francisco and the swellest social-athletic club of Oakland."

Fingers Greasy? Use Your Breeches!

The Army's door was open. The food was good. Everybody was welcome. Who cared if manners were not always quite so swell? "Napkins be blowed!" yelled a young gentleman from the South of Market's Tar Flat in 1895 when someone complained there were no napkins for greasy fingers. "What's the matter with your breeches?"[2]

Remembering the Basement

Table manners probably weren't much improved in the 1960s and early 1970s food lines, where serving was still pretty primitive. Harbor Light Advisory Council member Ernie Marx remembers when holiday meals for Skid Row were cooked and dished up in the basement on Fourth Street, then passed hand over hand to volunteers up the stairs and into the crowded dining room where scruffy alcoholics and other lonely people, mostly men, stood at high tables to eat. The streets were bleak. To an outsider, the atmosphere inside seemed pretty bleak as well.

But standing at the platform was Nick Gabriel telling his story, "One year I was with my family at home enjoying Thanksgiving. Then I began celebrating alone in a bar. In other years I was on Skid Row, alone and drunk. Finally, I came to Harbor Light . . ." Gabriel eventually served as director of the center.

Not everybody who came into the building had spent the night in the gutter. Periodically, there were men in hats, suits and ties, carrying the morning newspaper tucked underneath their arm . . . cuffs frayed, not quite down and out, but still on the dole.

"When the Saints Go Marching In"

When Harbor Light moved to the Gladding McBean building at Ninth and Harrison in 1972, holiday cooking was done in a kitchen off the main building. Volunteers waited in the courtyard to get their routes for deliveries to seniors and disabled. After home deliveries were taken care of, staff then delivered huge pans of turkey and dressing, vegetables and cakes for the homeless to a warehouse space at 125 Valencia. More than a hundred volunteers handed plates forward from the serving tables. It was crowded. There was barely room to walk between the tables. It was a warehouse There was no denying that. But the feeling was upbeat.

The Bay Street Jazz Band performed and always ended with a hand-clapping, around-the-room parade, a rousing rendition of "When the Saints Go Marching In." When people left, fresh fruit in hand, they were smiling.

1980s and The Working Poor

During the 1980s, people who came to these dinners were still scruffy and still mostly men. But it gradually became apparent there was a perceptible age difference in those who came to eat their meals at the "Sally."

They were younger.

They were Vietnam vets with drug problems . . . leftover hippies . . . worn out and vulnerable people who had been living on the streets since the closing of California State mental hospitals . . . the working poor, low-income people who simply couldn't make ends meet. And each year, all the spokespeople for meal programs throughout the city said by rote, "We saw more families and children this year."

Why Duplicate?

In the mid-1980s, The Salvation Army made a decision. Word on the street was that homeless people were enjoying the meals of as many agencies as possible on the holidays. Nobody begrudged them that. But The Salvation Army likes to stretch a dollar as far as it'll go. Everybody knows that. As more agencies began serving meals, it seemed that rather than duplicate services, it would be better to reach out in new directions, to do more about the underlying need.

All the holiday meals in the world—no matter how welcome to a hungry family—are just the beginning of service to homeless people. They meet an immediate need, but not the underlying root of the problem. Christmas, after all, is just one day out of the year. What happens the other 364 days?

It was during this period that The Salvation Army made a significant shift away from soup kitchens and holiday mass feedings. Nationally, the organization had made a similar decision in the 1920s when Commander Evangeline Booth called a halt to huge Christmas dinners in Madison Square Garden and began an annual distribution of over 500,000 food baskets for families.

Using donated food and canned goods plus USDA commodity food and grocery store discounts, the Army continues to distribute food boxes at neighborhood centers weekly and at Thanksgiving and Christmas. Depending on what's available, staples include canned vegetables, rice, cereal, peanut butter, fresh vegetables and a Safeway voucher for holiday ham or turkey.

The Army canceled its sit-down holiday meal for street people and expanded holiday food distribution to the "hidden poor"—seniors and disabled who live alone in single room occupancy downtown hotels and those who live without family or friends on limited incomes in the avenues, people for whom the holidays are often very lonely days.

More than a dozen agencies now ask the Army to take meals to their clients at Thanksgiving and again at Christmas.

Carving the Birds

It takes a hundred birds to feed 3,000 seniors at Thanksgiving. How they get roasted, carved and served is thanks to kitchens like the USS Carl Vinson, the Westin St. Francis, the Marriott and the Fairmont, plus people like the carving knife brigade from the SF Fire and Police Departments, the US Marine Corps and Golden Gate Breakfast Club.

Harbor Light council member Ernie Marx holds the title as all time lead carver.

Holiday Mornings on the Streets of San Francisco

The streets of San Francisco are surrealistically quiet at 8 a.m. on Thanksgiving and Christmas mornings. You can drive downtown barely passing a car, gliding through intersections and believing you must indeed be the only person out on the road.

And suddenly, at the corner of Ninth and Harrison, there's a double line of cars in the middle of the street waiting for 300 volunteers to bring out almost 3,000 hot holiday meals. Inside, a process that in its earlier years was often pure chaos has become so organized that volunteers are welcomed, assigned to computerized routes and out the door within an hour.

Recovering alcoholics and drug addicts make it happen. This highly charged, highly personal approach to being on the giving side of the holidays is an important part of the program. For many in recovery, this is their first time in years to celebrate the holidays with a family which, although they may not be blood relatives, are a community of peers.

Surrounded by an army of volunteers and merrily dishing out meat, vegetables and gravy, the kitchen serving crew was belting out "Lean on Me" when the

TV cameras arrived. A young addict smiled and said volunteering was addictive. She eventually graduated from the program and was on her own for three years—better for a long time, then suddenly worse. She died of an overdose.

"I Used to Be a Donor"

A man, perhaps 70 years old, called on Christmas Day to ask if The Salvation Army could bring him a meal. When a volunteer arrived at his house with a dinner of ham, green beans, sweet potatoes and oranges, he came to the door wearing pajamas and a heavy growth of beard.

It seemed a genuinely embarrassing moment for him. He apologized profusely, saying that he had in fact been a Salvation Army donor in the past.

Lately, he said, during months when his wife had been in the hospital, he had become extremely depressed. Everything he had worked for "seemed to be falling apart."

On Thanksgiving, at the hospital with his wife, he had forgotten to eat at all. On Christmas, his only visitor was The Salvation Army.

The Personal Touch

No matter how high tech the world becomes, Salvationists maintain that it's the personal touch that makes the difference in people's lives. Each year—throughout the year, but especially during the holidays—thousands of hospital patients, seniors and invalids receive friendly visits from Army League of Mercy volunteers. It's a "ministry of presence"—that's the personal part, a human being taking time to visit. It's also practical, with a warm pair of socks thrown in, toiletries, stationery and holiday ornaments, along with multiple hugs.

A Unique Distribution Center

The Salvation Army Service Center at 520 Jessie—which becomes an infant and toddlers' clothing distribution center in the spring and a students' "back to school" clothing and supply center in September—at Christmas becomes a toy store.

The Army's great strength is that what it receives, it finds a place for and a way to distribute. Shipments of clothing, food, toys, teddy bears and every kind of excess inventory item imaginable find their way to the Jessie Street Service Center throughout the year. Toys are distributed the week before Christmas.

No other agency does it quite the same.

City units sign up families in advance and carefully make sure it won't be a mass giveaway. It's not first-come, first served. Everyone has a chance to shop.

There are no lines outside the door. Distribution is by prior appointment – parents only, no children allowed. Parents are invited to select gifts for their children and when they take the packages home, there's no indication they came from The Salvation Army. Unless children ask, there is no need for them to know.

It's still charity, of course, but, at its best, the system tries to offer some sense of dignity to parents selecting gifts for their children.

Divisional Headquarters staff, Advisory Board members, employee groups, churches, business organizations and parents help center staff for this operation. However you look at it, somebody has to physically move the goods, organize the toys by age group and re-stock the tables. Warehousing is an essential ingredient in most Army operations. Everybody takes a turn.

When It's Your Own Family

Christmas is a tricky time for people at the Army, a time when—just like the rest of the world—they can easily get caught up in the rush of the holidays.

Salvationists and Army employees can remember years when last minute gifts were bought at midnight Christmas Eve, when the family tree didn't get put up until Christmas afternoon, when the phone continued to ring even as the turkey was coming out of the oven.

They also remember precious times of caroling together with family and friends for seniors at Silvercrest, taking gifts to the Veterans' Hospital and to patients at the AIDS ward at San Francisco General. In many Army homes, officers teach their children to volunteer in the morning before going home to open their gifts.

As with everything else, approaching the holidays is a balancing act. People who sign on for service know the holiday season at The Salvation Army is an incredibly challenging experience.

Two days before Thanksgiving, a crew of officers and employees deck the halls. Lights recycled from the

Symphony Christmas tree of the year before surround every door. Wreaths are on the walls, a manger is in the lobby and an entire construction-paper starry night stretches across one end of the dining room.

Phones are ringing. The malls are worried again about Christmas kettles. Will on-line fund raising be the wave of the future? There are new-fangled ways of looking at the season . . . but some things never change. The band practices down the hall. Carolers sing. There are lots of parties. Bell ringers ring.

At chapel, Advent candles glow brightly. This is a religious holiday. Above all, the Army remembers this. Through all the busy-ness, and of course, because of it, the Christmas message rings through with wonderful clarity.

Silver Bells!
A Seventeen Year Old Rang the First One

It was winter, 1901. The weather in New York's Bowery District was freezing and the wind was fierce. Seventeen-year-old Amelia Kunkle stood at her Salvation Army kettle and couldn't, for the life of her, get the attention of people passing her by.

The kettle tradition, which had begun in San Francisco and slowly spread east, was the Army's prime Christmas fundraising. On this day, it simply wasn't working.

Exasperated, Amelia walked into a dime store and bought a bell. Yes, she said later, it did drive the employees crazy in nearby bank buildings, so she bought a smaller one. But it also proved a point. When she rang the bell, people turned around to look. And if they stopped long enough to listen, they usually put their hand in a pocket to pull out a donation.

Lore has it that at least once John D. Rockefeller coughed up a twenty dollar bill for a Wall Street kettle—a significant chunk of pocket change in those days.

As the years went by and the bells continued to ring, shoppers came to associate the bell with the holiday season. Bob Hope sang "Silver Bells" written by Ray Evans and Jay Livingston in a 1951 Paramount Pictures movie called *The Lemon Drop Kid* and the song has since become a Christmas tradition. It was not written expressly for the Army, but it's an obvious connection.

Silver bells have become a symbol of The Salvation Army's Christmas effort.

What a lot of fuss over a tiny bell, said the first bell ringer in numerous newspaper articles over the years. She became a celebrity of sorts, not only for the bell, but for a life story that included surviving the 1889 Johnstown Flood which killed 2,200 people and working at age 12 and 13 to help support eleven siblings. Amelia served in 13 Salvation Army appointments.

Her last assignment brought her to San Francisco's Barbary Coast. And it was here that she fell in love and married Andy Devine. Andy was a singer with The Salvation Army, but not an officer. Amelia gave up her commission and concentrated on raising a family.

After her husband died in 1929, she cleaned houses and took care of babies to put their four children through college. Working on the outside, but still a soldier in The Salvation Army, she lived in a Burlingame apartment for fifty years before moving to Grover City, California, where she died in 1984 at age 100.[3]

Guys and Dolls
The Movie and the Reality

Who was that slick-talking, deal-making, marriage-shy weasel of a gambler in a double breasted suit Nathan Detroit? Frank Sinatra, of course.

In the 1955 highly-stylized film musical *Guys and Dolls*, which is probably how most of the world knows him, Nathan Detroit bragged that he had run the "oldest established permanent crap game in New York since the days when he was a juvenile delinquent." Adroitly, he slithered in and out of a 14-year commitment to marry his long-time sweetheart Miss Adelaide (Vivian Blaine). She, in turn, sneezed woefully every time she thought about the delay.

Smooth as silk, singing for the only time in his career and always ready to take on a bet, Sky Masterson (Marlon Brando) wooed and surprised even himself by winning the prim and very proper Sister Sarah Brown (Jean Simmons), an evangelist at an inner city mission.

Loosely, very loosely, her character was modeled after a Salvation Army lassie.

Finally, Nicely Nicely—the affable sidekick played by Stubby Kaye—brought the plot to a sudden change of heart in a memorable midnight prayer meeting attended by thugs and otherwise lovable gangsters. "Sit down, Sit down. You're rocking the boat!" he sang.

And the angels smiled. Nicely Nicely had seen the light.

All this was going on in the streets of New York. The time was after hours. Way after hours.

A Real Lady Rocked the Boat

It's generally thought that Salvation Army Captain Rheba Crawford was the inspiration for Sister Sarah. She was a dramatic young captain whose name for a short time was also associated with Aimee Semple McPherson. In the 1920s she brought huge crowds of people to midnight open-air meetings in Times Square. She was called "The Angel of Broadway."

They say as many as five and six thousand people a night came out to hear her. One night she was arrested for obstructing traffic, and the crowds were so distressed they crashed through the door of a local police station to protest. Columnist Walter Winchell heard about it and rushed from the Palace Theater with a former police captain to release her. It took a little a bit of negotiation, but eventually the charge was changed to disturbing the peace. Damon Runyon got wind of the incident and wrote it up as a short story featuring Sister Sarah. So the story goes . . .

Damon Runyon's fictional characters were always caricatures; one-dimensional figures playing fanciful roles in a romanticized world. Big Julie, Harry the Horse, Society Max—they were small time hoods scrubbed clean for the silver screen and matched head to head with the virtuous lady from the mission.

It was Runyon's world, at least the way he imagined it for the reading public of the Depression years.

Biographies say Runyon was a recovering alcoholic who evidently stopped drinking cold turkey in his thirties, realizing his body simply couldn't tolerate liquor. He presumably never touched another drop, even though his newspaper beat and lifestyle often revolved around speakeasies and underworld crime.

His father had been a heavy drinker, leaving Damon pretty much to grow up on his own on the street. The story is that at one time they had only one bed and one pair of shoes between them. They used them in shifts. On Christmas Day, they stood in line to eat dinner with other loners in a local saloon.

In 1898, barely old enough to enlist, Runyon passed through San Francisco on his way to military service in Manila, where he was stationed in the city's red light district. According to Michael D. McClanahan of the *Denver Post*, Runyon blew his final paycheck in one wild week in San Francisco, and then rode the rails back home.

Years later, estranged from his father but trying to reestablish contact, Damon Jr. also admitted to having had a hard time with alcohol. His father had praised Alcoholics Anonymous. Did he ever have more than a passing contact with The Salvation Army? Runyon's life was full of the kinds of characters and situations that bring people to The Salvation Army. Maybe he never knew the Army personally. But he did know about alcohol. And he knew about the Bowery days of the Roaring Twenties.[11]

In an endearing, unforgettable way, he immortalized the innocent officer, Sarah Brown.

The image of Sister Sarah in the Save a Soul Mission is the only picture some people have of The Salvation Army. It's a gentle, benevolent, old-fashioned image. Salvationists laugh at the story as hard as anyone else.

Theater groups all over the world produce the play. Production managers call often to ask the Army for help with costuming the performance. However, Salvationists pay for their own uniforms and "wear them forever," so school costume departments aren't likely to find used uniforms. During the 1980s, the Army produced a number of theatrical performances based on the group's history. Some costumes from these shows still exist and are loaned out for local productions.

Costumes for the play and later the 1955 Technicolor movie were modeled after the high collared Salvation Army uniform. Inspiration for the burgundy color came from the piping used on regulation Army uniforms. If you look at the top of the women's bonnets, you can see the resemblance to those worn in the Army's early years.

Damon Runyon had churned out 1930s tales about the antics of Sarah Brown, Apple Annie and Dave the Dude for readers of *Colliers* and *Cosmopolitan*. Jo Swerling and Abe Burrows compiled them and after Runyon's death, Frank Loesser, Cy Feuer, Ernest Martin and Abe Burrows adapted "The Idyll of Miss Sarah Brown" for the stage, borrowing the crap shooting scene from an earlier story. The play opened at the 46th Street Theater on November 24, 1950, and ran for 1,200 performances, netting $12 million. It was one of Broadway's greatest successes at that time.

Major productions of the story have included a Hollywood version with Milton Berle . . . a Broadway revival in 1976 . . . and a 1992 Broadway revival with Nathan Detroit portrayed by Nathan Lane, who actually took his name from the character and won a Tony for his portrayal.

Major Barbara

George Bernard Shaw's *Major Barbara*, of course, is the other major theatrical production which colors people's perceptions of the Army. Shaw wrote the play in 1905 and it opened in the Court Theatre in London on November 28, 1905. He attended the opening, along with much of London's literary crowd. Still cautious of worldly pleasures inherent in the wicked stage, the Booth family did not. However, they must have tacitly approved the production, since several Salvation Army officers were in the audience.

Grace George took the title role in the first American production which opened at the Playhouse Theatre in 1915. Wendy Hiller, Rex Harrison and Robert Morley starred in the 1941 movie. Glynis Johns, Eli Wallach and Charles Laughton starred in a 1956 production. A Caedmon Production by the Theatre Recording Society features the voices of Maggie Smith and Robert Morley.

In the 2004 production at the San Jose Repertory Theatre, Major Sheryl Tollerud representing the Army in Santa Clara got into the act. As a consultant, she found Army uniforms for the production, banners and even a local recording of the "West Ham March."

Shaw questioned whether a charitable organization should accept a donation from a munitions manufacturer who made his money arming countries for war. William Booth always had an answer for a question like that. When chastised for accepting a donation from the Marquis of Queensbury, a professed agnostic, he said thank you very much and went ahead to see what good he could make of it.

In the booming, oratorical voice we've come to recognize, he determined, "We will wash it in the tears of the widows and orphans and lay it on the altar of humanity."

Tainted Money

For years—and only half in jest—old time Salvationists often had a quick response to the concern about so-called tainted money. When the economy sags and times are lean, they said, "Even a good donation 't'ain't enough!"

Show Business

Is it a coincidence that The Salvation Army has a closet full of costumes and a brass band willing to play at the drop of a hat? Hardly. William Booth and his progeny were consummate show people.

William Booth jumped on a chair and waved a handkerchief wildly to illustrate the distress signal of sinners shipwrecked in an ocean of evil. Bramwell Booth jumped out of a coffin to make his point about eternal life. Dressed in rags, Evangeline Booth mingled with East End crowds in London and later, in similar rags, developed a popular elocution number in America that left audiences weeping and reaching for their pocketbook to help the poor. To get the crowd's attention, James Dowdle, the Hallelujah Fiddler evangelist, plopped his fiddle case on the sidewalk and warned, "Stand Back! It might go off." Eyes wide, the audience watched as he pulled out his fiddle and played a jig.

Early Salvationists evidently went to great lengths to draw attention to themselves on the streets of London. Would they wax so dramatic today? Not usually. But deep down in their hearts, and with great fondness, Salvationists often remember the antics of their forebears. Many of them love dressing in costume, play-acting and reenacting the olden days.

The 1970s and 1980s were a particularly creative time for local Salvationists. People remember young girls wearing candles on their heads to celebrate the Santa Lucia Festivals in the auditorium at 101 Valencia, a tradition started by the Scandinavian waterfront corps. They remember "Thursday Night at the Army" when visiting brass bands and trumpet soloists came regularly to perform. They took Army productions "on the road" to the Lamplight Theater on Fillmore and even to Southern California. For a Calypso Christmas performance by the Training School, Principal Harry Larsen sat on the floor dressed in bright calypso colors and beat a calypso drum.

For the annual meeting honoring the San Francisco Forty-Niners in 1982, officers re-enacted "soup kitchens" in each corner of the Fairmont's elegant Grand Ballroom. When Justice Sandra Day O'Connor was honored in 1984, a group donned the costumes of "Slum Sisters" who lived and worked in the slums of England and America.

Finally, during the 1980s the fortuitous pairing of John Gowans and John Larsson created a musical team that in The Salvation Army world might as well have been Rodgers and Hart.

"Take Over Bid," "Hosea," "Jesus Folk," "Spirit," "Glory!," "White Rose," "The Blood of the Lamb," "Son of Man," "Man Mark II," and "The Meeting" were popular musical performances based on Salvation Army history and belief. The shows were performed in the United States, the United Kingdom, Australia, New Zealand, Japan and Canada. Local officers and soldiers, grandparents, teens, children, friends and neighbors took lead roles as well as bit parts. Barry Frost directed each one of the plays at least once.

The Army's Northern California and Nevada Division produced "Blood of the Lamb" for several thousand people at the Army's American Centennial Congress in Kansas City in 1980.

In San Francisco as far back as 1948, a radio production began that continued through the 1980s and was aired annually by over 1,000 radio stations throughout the United States and Canada. It was the popular "Army of Stars" series. Using performers from the San Francisco Opera Company, Lloyd Docter created a production of professionally performed Christmas music. In the early years, opera stars like Ezio Pinza, Kirsten Flagstad, Lily Pons and Jan Pearce donated their services. In later years, Dr. Robert Docter continued in his father's footsteps by asking people like Vincent Price, Angela Lansbury, Walter Cronkite, and James Earl Jones to narrate. Additionally, the Mutual Broadcasting System annually aired his dramatic story of Christ's birth titled *Thou Bethlehem*.

In 1955 Docter also wrote *Heartbeat Theater*, a radio program to illustrate the drama of the Army's ministry which ran in nearly 100 stations and in the Armed Forces Overseas Radio system. Fifty years later, it's still heard on over 300 broadcasting outlets worldwide, as well as on the Internet.

Half Time Extravaganzas

In recent years, thanks to National Salvation Army Advisory Board member Jerry Jones and the Dallas Cowboys, the Army has kicked off its Christmas Red Kettle Campaign during the halftime extravaganza at the Thanksgiving Day football game.

Nationally televised and extravagantly staged, the shows have included headliners Reba McIntyre, Randy Travis, Creed and Beyonce Knowles, Kelly Rowland and Michelle Williams of Destiny's Child.

At the Turn of the Century
Productions Like Salvation Nell *Idealized The Salvation Army Lassie*

The theater was sold out. Salvation Army lassies passed the hat during intermission. Beyond the proscenium arch, storm clouds threatened and sounds of thunder echoed ominously as the melodrama unfolded before an enthusiastic audience.

That's the heart-wrenching pathos Edward Sheldon (1886-1946) included in his 1908 play called *Salvation Nell*. Theater critics Clive Barnes and John Gassner included it in the 1969 anthology, "50 Best Plays of the American Theatre," not because it was such a great play, but because it so realistically described the times.

The production was hopelessly romantic—the story of a scrubwoman who joins The Salvation Army and forgives her man again and again as he drunkenly defies the law. It played out in realistic detail street scenes of early day slums.

The play opened at the Opera House in Providence, Rhode Island, on November 12, 1908. Five days later, it opened at the Hackett Theater in New York. Mrs. Minnie Maddern Fisk starred. Film versions were produced in 1915, 1921 and 1931.

Goodness and Light

Salvation Army lassies epitomized goodness and light in theatrical performances at the turn of the 20th century. These were just a few of the titles that captured and sometimes recaptured theatergoers' hearts.

The Belle of Broadway (on stage in 1897, on screen with Marian Davies in 1919, revived with Vera-Ellen and Fred Astaire in 1952) . . . *The Chorus Girl and The Salvation Army Lassie* (1903) . . . *Salvation Lass* (by D. W. Griffith, 1908) . . . *Fires of Faith* (1909) . . . *The Little Girl That He Forgot* (1915) . . . *A Gutter Magdalene* (1915) . . . *Salvation Joan* (1916) . . . *How Could You, Jean?* (with Mary Pickford, 1918) . . . *Salvation Rose* (1919) . . . *The Blue Bonnet* (1919) . . . *Hell's Oasis* (1920) . . . *Human Wreckage* (1923) . . . *Salvation Jane* (1927) . . . *The Angel of Broadway* (1927) . . . *A Goodbye-Kiss* (a Max Sennett production, 1928) . . . *The Street of Sin* (1928) . . . *Laughing Sinners* (with Joan Crawford and Clark Gable, 1931) . . . *The Miracle Woman* (1931) . . . *The Law of the Tong* (1931) . . . *She Done Him Wrong* (with Mae West and Cary Grant, 1933) . . . *Limehouse Blues* (1934) . . . *Smashing the Vice Trust* (1937) . . . *Hello, Frisco* (1943) . . . *Abandoned* (1949).

Research for over 100 films—historic and contemporary—with Salvation Army references, was originally done for a videotape presentation prepared by Barry Frost and Ken Bricknell for showing at a 1999 advisory organization conference in Pasadena. They, plus Carla Gerard from the Army's Western Territorial Headquarters and film editor/director/historian Rich Henderson, have continued to update the material. Additional information is available through the THQ community relations department. Their research starts with Australia, where as early as 1891 Salvationists were far ahead of the times in the motion picture business. In an abandoned attic, lassies hand painted glass slides to illustrate General Booth's "Darkest England" book tour, improvising equipment to make movies and sound recordings while filming everything in the view finder's sight. The Australian government credits the Army's Limelight Department with much of its historic footage, including the opening of the Federal Parliament in Melbourne in May 1901. By that year, the Army also had 270 hand painted slides to project in concert halls or on the side of a building in the square.

Today, whenever The Salvation Army shows up in a movie—no matter how fleetingly—there's a small cadre of Salvationists somewhere taking note. It seems a small thing, but considering how much money some organizations put out to make sure their image is on screen, it's nice to note that The Salvation Army doesn't pay for what advertisers call "product placement." (an organization paying for its logo to be in the picture.)

In the early years, before and just after World War I, when doughnuts were fresh on people's minds, Salvation Army slum workers were often part of the story line.

Today, the bands are simply such a ubiquitous part of the urban landscape you don't think to notice them until they're gone.

A Reassuring Sound on Mean City Streets
"Three Days of the Condor"

An ingenuous and increasingly disenchanted Robert Redford hurtles through New York City streets frantically avoiding multiple disasters as he tries to decipher who the good guys are, who the bad. Suddenly, in the background, Director Sidney Pollack interjects a reassuring Salvation Army band led by Russell Prince (who served as Development Director in San Francisco during the 1980s) playing an old standard. The image is taken for granted on busy city streets and provides a sharp "business as usual" counterpoint to the mounting intrigue.

And The Band Plays On
"North to Alaska"

Mud fight! Seattle in the 1890s is mired in mud as the movie rushes to a free-for-all climax that captures the fighting spirit of the Gold Rush. Ankle-deep in the mud, John Wayne and Capuchine wrestle with law and order, love and plot development. Unflappable, The Salvation Army band plays on . . .

Notes

[1] *The San Francisco Morning Call*, December 26, 1891

[2] *The Conqueror*, 1895, page 69

[3] Newspaper interviews with Carolyn Young, *San Francisco Examiner*, August 20, 1980; Doris Bentley, *Telegram–Tribune*, San Luis Obispo County, March 2, 1984; Associated Press obituary, April 28, 1984

[4] Bernice Sharlach, "Dealing from the Heart, A Biography of Benjamin Swig" Scottwall Associates, Publishers, San Francisco 2000

[5] Herb Caen, *San Francisco Chronicle*, December 19, 1982

[6] "Stuff That Makes An Army," Colonel William Harris, 1962, Page 102

[7] William Booth in his first "Order for Bands" from *The War Cry*, 1880

[8] "Beginnings of The Salvation Army in San Francisco" written by Major Alfred Wells in 1925

[9] William McKinley attributes this quote to "Salvation Limited," an article in *The Nation*, January 30, 1929, p. 24.

[10] Incident drawn from "The California Venture, The History of The Salvation Army in the West, 1883–1982," by Mary A NeHalsingh, Torrance Corps.

[11] For further reading: Runyon, Damon, Jr. "Father's Footsteps," New York, Random House, 1954; Breslin, Jimmy, "A Life of Damon Runyon," Ticknor & Fields, New York, 1991; Mosedale, John, "The Men Who Invented Broadway, Damon Runyon, Walter Winchell & Their World." Richard Murek Publishers, New York, 1981; Internet sites, "The Story of Damon Runyon printed by the Denver Press Club Online, copyright by Michael D. McClanahan, 1999.

ABOVE: *The look may have changed, but not the commitment.*

Christmas Cry the Very Next Issue!

WAR CRY

Official Gazette of the Salvation Army in the United States.

WILLIAM BOOTH, General.   Entered at New York Post Office as Second-Class Mail Matter   F. and E. BOOTH TUCKER, Commanders

No. 1166.   120-124 W. FOURTEENTH ST., NEW YORK CITY, SATURDAY, DECEMBER 13, 1902.   Price 5 Cents.

KEEP THE POT BOILING.

SEVEN

South of Market

Changing the Skyline

If you stand today in The Salvation Army's 832 Folsom Street Divisional Headquarters and look South, you can almost count the Army's history in square feet.

From the day Major Alfred Wells landed in San Francisco in July, 1883, and settled into a boarding house at Fourth and Mission Street, the South of Market (SoMa) was Salvation Army territory.

There's a long history of Salvation Army soup kitchens and alcoholism treatment in the area. For years, the Army has invested in real estate here, and in people.

Architectural historian Anne B. Bloomfield has recorded that there was an increase in area charities as a result of the Depression of 1893. By that time, The Salvation Army had a corps opposite the Grand Opera House, a children's home on Second Street, and a receiving home on Jessie. "These didn't last long," she said. "But the wood yard remained, where out-of-work men could chop wood to earn a meal or a bed."[1]

In 1897 the Army kept care of 31 children in a home originally designed for a well-to-do San Franciscan at 328 Bryant across from St. Mary's Hospital.[2]

At 826 Folsom that year, a relief store distributed 3,418 garments in four months. At 829 Folsom was a shelter for women. Check the address. The Army's Golden State Divisional Headquarters, purchased many years later, is located on the same block—at 832 Folsom.

Homelessness in 1924

A 1924 Community Chest Study on Homelessness in San Francisco painted a sorry picture of the district lying between Second and Sixth Streets and Mission and Harrison. With it was a warning to those looking for quick solutions.

"The influences that make for homelessness and vagrancy growing out of it are nation-wide in their ramifications, and have their roots deep in defective social and industrial conditions which will not readily yield to local treatment," said the report.

"So any loud claim of a radical cure for this social malady by isolated cities or communities may be wisely taken with a grain of allowance."[3]

The study covered The Salvation Army wood yard at 876 Harrison, which burned in 1923, The Salvation Army Relief Office at 243 Golden Gate and the Volunteers of America wood yard at 119-10th Street.

The results were predictable. According to the report, most of the people served were "old and virtually unemployable."

However, there was a common belief that these were transients, men always on the move from one city to the next. The study showed exactly the opposite. Two out of three had a regular residence of some sort in San Francisco. Half were registered to vote locally.

There evidently were excellent free clinics in the

area, but agencies like the Army had no field workers or nurses. As much as they tried to help, assistance was temporary at best.

Most agencies used the Standard Hotel (995 Folsom) at Sixth to house their homeless men. It apparently was the only hotel that would care them.

In later years, a man remembers his dad warning him as a teenager what a vile fate would await him if he didn't mend his ways. "You'd better shape up, or you'll end up down on Third Street with The Salvation Army," he said.

And it was true. For years the South of Market was a vast wasteland of lost souls and dirty streets. "We saw it from the freeway, but we never went down there," remembers another. "We just hurried through the Third Street and Fourth Street corridors to get from the Southern Pacific Station to work in the Financial District in the mornings and through it again to get back to the commuter train at night."

The Army was in the area before Redevelopment and dot-coms, high rises, trendy restaurants, nightclubs and live-work lofts.

It's never left.

Harbor Light services for street people and alcoholics moved from Hunt Street to Fourth Street and eventually to Ninth and Harrison in 1972.

In the 1970s, the Army made another major investment in the neighborhood.

It went into the senior housing market.

More Than a Landlord

"The Salvation Army should be much more than just a landlord," said Lt. Colonel Ray Robinson as he looked at the Army's twin ten-story high rises in the South of Market.

The buildings stood out boldly against a skyline of broken down Skid Row hotels.

Robinson had envisioned and nurtured the Silvercrest Housing for Seniors project years before when he was General Secretary. When it opened in 1974 under the leadership of Lt. Colonel Victor Newbould, the first residents were an economic, geographic and ethnic mix.

Some were from Single Room Occupancy hotels in the North of Market.

Some were indigents displaced from firetrap hotels in the

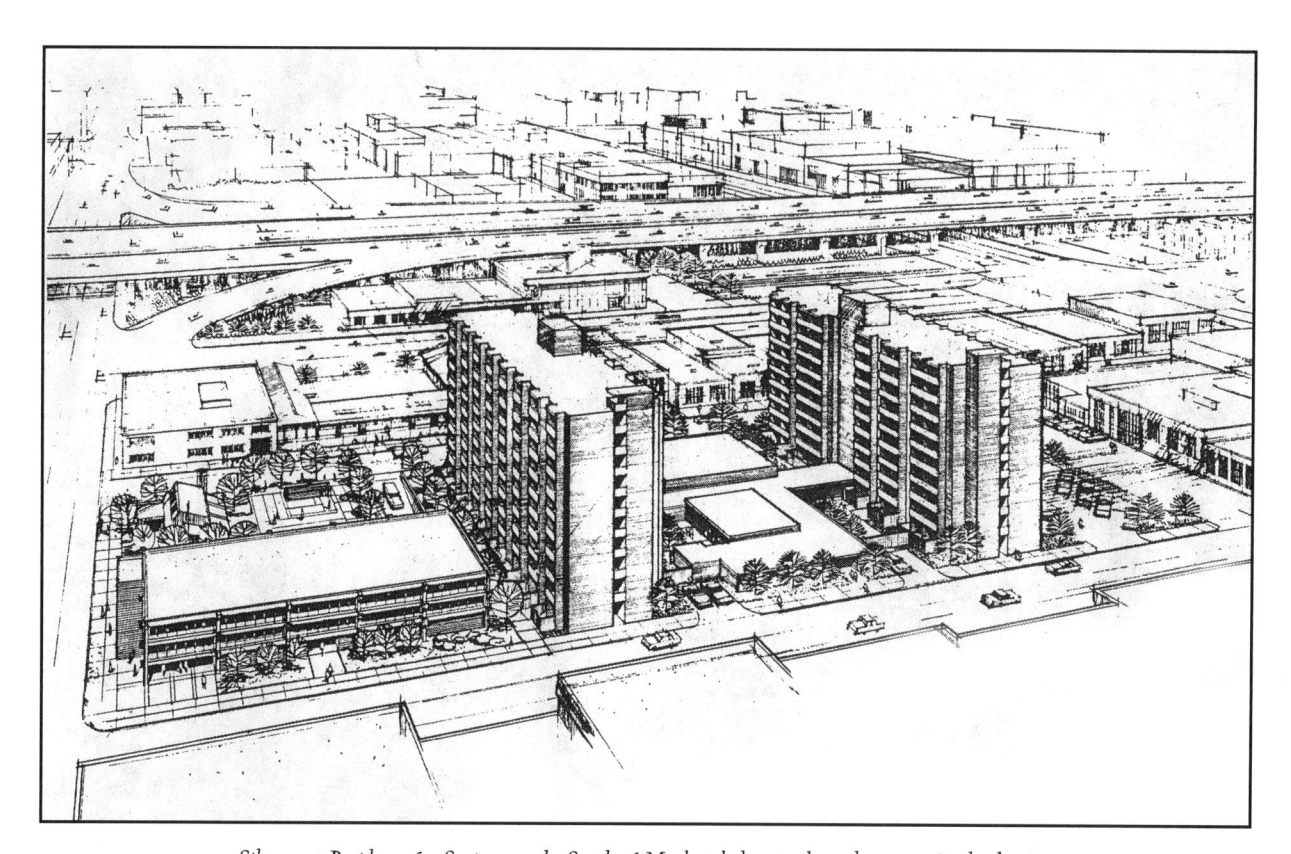

Silvercrest Residence for Seniors on the South of Market skyline in the early years of redevelopment

South of Market. Others, paying market rate, moved from homes in the Avenues they could no longer afford to keep.

For the most part, the neighborhood they came to was bleak.

There were few hints of the redevelopment that's there today. Directors Clinton Steele and, later, Gene Lantz and Ken Iremonger had their work cut out for them.

Since Silvercrest is so near San Francisco's original shoreline, it sustained damage during the Loma Prieta earthquake in 1989. In 2004, Lt. Colonels Richard and Bettie Love initiated an extensive remodeling, utilizing low-income tax credits, tax-exempt bonds and Federal Home Loan Bank Affordable Housing program funding.

A kitchen and dining room connect the two buildings. Originally it was built to serve the 257 tenants, but that was before Virginia Wolf and Evie Dexter began to explore the concept of a Senior Meals and Nutrition Program. Ultimately the program served over 465,000 meals a year at 33 meal sites. The City contract was canceled in 1997 during negotiations over Domestic Partners legislation.

Administration for the program was housed in the Senior Activities Center next door and, under Dexter's leadership, it was an exciting building. What had been an old Police Station at 360 Fourth Street and then a soup kitchen and halfway house for alcoholics was now a hive of committee meetings, activities and concern for making the neighborhood safe and invigorating for seniors. Something was happening in every room.

Staff was bright and loving. They made it clear they were there to do things *with* seniors not *for* them. With them came a spirit of independence and empowerment and assurance that growing older should be positive and creative. You could feel it when you walked in the door. It was part of Dexter's management style. "I was always so proud to share with staff what the Army was all about," she says.

When the nutrition program moved to 850 Harrison, the building became the Central Corps. It was renamed the Yerba Buena Corps to strengthen its relationship to the neighborhood. Still later, it became the Yerba Buena Asian American Corps.

A Really Big Kitchen

When the central kitchen opened at 850 Harrison it was heralded as the largest Salvation Army kitchen in the country, a claim yet to be challenged. If absolutely necessary, pushed to its limit, Mike Afshar and the kitchen staff could presumably churn out 30,000 meals in 24 hours. Fortunately it's a claim that's never had to be tested.

The new kitchen had 16 ovens, 10 stoves, two grills, 80-gallon soup vats and 3,760 cubic feet of freezer and refrigerator space.

These were the same years the Senior Meals Program was putting on huge picnics in the park. On one day of the year, instead of delivering meals to the sites, the Army reversed the process and delivered nearly 2,200 seniors to the park. Originally, the party was at Stern Grove. Later, it was moved to the concourse in Golden Gate Park. For seniors and staff and volunteers, it was always a remarkable, amazingly sun-filled day.

Not once in its 21 year history did it rain.

During this time, the program secured funds to expand a service that had been attached to the center since 1985. A Wednesday morning round with the Sheriff's Eviction Assistance Program (EASE) was eye-opening. Sgt. Frank Hutchins and his crew met with seniors and disabled about to be put out on the streets. They knocked on doors of end-stage alcoholic seniors living in filth. Salvation Army representatives—seniors, themselves—were there to help in whatever way they could.

Yet another program, "Meals That Heal," continues still and concentrates on delivering a hot, nutritionally-balanced meal daily to seniors living alone in single-room occupancy hotels at Silvercrest and in the Tenderloin. Besides the meal, a major part of the assignment is simply to check on each person each day. Often, these are the only visitors a senior sees. They're greeted warmly.

A tiny, wizened woman with a sweet smile introduces volunteers to her cat, her only companion.

A man with a night table full of prescriptions sits on the side of his bed when lunch arrives.

A man in a hospital bed pulls himself up to welcome guests.

Reaching Out

"In 1984, the Army expanded its outreach through Relief for Energy Assistance through Community Help (REACH). Through corps and service extension units, the Northern California and Nevada Division literally touched every community within PG&E's range. By linking its units with PG&E, the Army was able to assist its regular clients. It also was a way of finding new ones.

***This architectural rendering** of Harbor House shows the outdoor play area next to the city's first licensed day care facility for homeless children. Harbor House (formerly called Gateway) at Ninth and Harrison gives homeless women in recovery with children a clean, safe residence in which they have up to two years to stabilize their lives.*

Is it urban legend or did it really happen? The story is often told that a Salvation Army worker went to a woman's home to help with her utility bill and opened the refrigerator to find nothing but a half-eaten can of dog food.

The woman, a frail senior, had no dog.

Army records are filled with such stories.

The New Poor

The press coined a new phrase in those days. "New Poor." Those were people who normally got by—just barely—but as the economy wavered, found themselves just a paycheck away from disaster and finally out of work.

"The President says he's 'perplexed' about the hunger situation," said a reporter over the phone. "Can you describe what's going on?"

Lines outside the Social Service Center around the corner from 101 Valencia illustrated the point.

Groceries were a much-needed commodity. They also became a great staple for building community service.

For nine years KRON-TV's partnership in the Food For Friends project energized food drives in schools and businesses throughout the Bay Area. Each year nearly half a million cans of food were collected. It was a labor-intensive and success-driven effort. Schools vied for producing the most results.

KRON's commitment to the project and to the kids

was proactive. The station, at that time an NBC affiliate, put its entire staff on the promotion. Weather broadcasts were shot live at high-achieving schools and businesses. Pep rallies and special events were held in malls, the Embarcadero Center and Pier 39. Partnerships were formed with community groups whenever and wherever possible.

Everybody worked. Everybody won.

When it rained in the early 1980s, Salvationists took vans to the streets to hand out blankets and hot soup. Harry De Ruyter served on the Coalition for the Homeless and when after-school activities were done for the day in the McCoppin Street Red Shield Youth Center, he supervised a shelter there.

Since 1926 the gym had been devoted to young people. During Jack Wolf's tenure, there was a full schedule of basketball leagues and other activities. After school tutorial programs and sports events filled the hall in the afternoons and weekends. But at night in these years, the cots came out and people from the streets had a place to sleep.

Was homelessness a problem that could ever be solved? Certainly nobody thought the poor wouldn't be with us always, but there was a resolve in those days which thought it might not always be such a permanent part of the landscape. Mayor Dianne Feinstein, who served on The Salvation Army advisory board, worried that San Francisco would become a magnet for homeless from across the country. Unfortunately, in many ways she was right.

Who would have believed the problem would ever reach the proportions it had by the turn of the twenty-first century?

The Army made plans to build a new family shelter.

Transitional Housing for the Homeless

The years under Lt. Colonels Bruce and Dorothy Harvey's leadership were years of major building and remodeling.

Harbor House (originally called Gateway Transitional Housing for Homeless Families) was welcomed by the city in 1991 and underwritten by a number of corporations. Sam Ginn and Pacific Bell led the way with a $200,000 grant plus a $200,000 challenge grant. Hayman Construction was first to meet the challenge with a $100,000 donation.)

The building at Ninth and Harrison was the City's first transitional housing facility with licensed day care. It was clean and laid out in clusters so families could share kitchen facilities but have the privacy of their own rooms. Referrals immediately came in from homeless-serving agencies throughout the City. Corporations looked for ways to help as they saw the advantages of giving homeless families up to two years to stabilize their lives.

Divisional Headquarters

Downtown San Francisco was heavily overbuilt during the late 1980s. Corporate offices were moving from the city to the suburbs. In 1989, just months before the Loma Prieta Earthquake, the Army moved its Divisional Headquarters to 832 Folsom, a building which had faced foreclosure and stood empty for want of a buyer. Colonel Harvey and Advisory Board members determined that the new building would be a way to maintain a downtown profile for the Army in a neighborhood where it had long had a presence. Ironically, it was on the same block where three early Salvation Army units had been.

Geographically, the Northern California and Nevada Division now reached from the Oregon border to the southern borders of San Luis Obispo, Kern and Inyo counties, east to Reno and Carson City and all of northern Nevada.

In order to better manage so many units, The Division was split in half in 1995 forming: 1) the Golden State Division which includes San Francisco and the peninsula down to the southern border of Monterey County and east to Kern and Inyo counties; and 2) the Del Oro Division which reaches to the Oregon State line and includes most of Nevada, except the Las Vegas area.

San Francisco, the City, took on its own administration, which in the next three years had three commanders—Captains David and Jeanne Bowler, Majors Kenneth and Dorothy Osbourn and Captains Robin and Teresaa Hu. Larry Hostetler served as administrator in 1997.

The Next Generation

When Lt. Colonels Richard and Bettie Love came to the city in 1997, it wasn't long before they began investing the Army's money into San Francisco young people—the next generation. They had specialized in youth activities throughout their Army career. It was natural they would move in this direction. In San Francisco they built up

The old Southern Police Station became a soup kitchen for Skid Row, then a halfway house for alcoholics, then a senior center and administrative hub of the Senior Meals and Nutrition Program. It is currently the Yerba Buena Asian American Corps/Senior Center.

the summer camp program and purchased the Jessica McClintock warehouse at 520 Jessie Street, turning it into a service center for youth and families. The building came on the market in 1998 and the Army made the purchase in time to move in that Christmas. The dress designer's building was perfect for Army use, right down to the clothes racks for a new back-to-school program and a Toy and Joy Shop at Christmas.

Perhaps the best way to understand the Service Center experience is to recreate a hot September Saturday afternoon in the City. The dichotomies of just a few short blocks illustrate how separately and differently San Franciscans live.

- At Powell and Market, a protest rally attracts a large crowd of tourists and shoppers;

- At a table in front of Emporio Armani, a shopper lunches on a $14 salad and across the street in front of Wells Fargo Bank, a lone banjo strummer entertains for spare change;

- At St. Patrick's church, a bride emerges with nine glamorous attendants to ride on a motorized cable car to her reception.

On this same day just a few blocks away in an alley off Sixth Street, The Salvation Army finishes a three-day distribution of free back-to-school supplies and clothes donated by places like Sears and the Gap.

It's a record number—1850 kids get new back to school outfits.

Outdoors in the harsh sunlight the realities of sidewalk drug trafficking and using are as bad as any street in San Francisco. It's a scary crush of humanity.

Inside, it's just the opposite.

Each family is welcomed. A lot of kids get hugged.

These are low-income families in one of the country's most expensive cities, trying to make ends meet. According to the U.S. Conference of Mayors Hunger and Homelessness Survey of 2004, "low income households spend up to 45% of their income on housing." In San Francisco, they run their households on little more than $800 a month or, if they're working, maybe as much as $1,000. They're grateful for charity, but the Army tries

SOUTH OF MARKET

not to call it that. The warehouse is set up in such a way that it seems almost like retail.

For parents, the operative word is "appointment." This is not a mass giveaway, first come, first served. Nobody waits in long lines. Nobody sits in the waiting room long. Every family gets a personal shopper and spends at least half an hour selecting outfits.

In preparation for the project, over 300 volunteers have helped sort and size the clothing. Teenage volunteers, wise in the ways of what's currently in style, eagerly show students where to find new clothes for their first day back at school.

Notes

[1] "A History of the California Historical Society's New Mission Street Neighborhood" in *California History*, a publication of the California Historical Society, Winter, 1995/96

[2] *Pacific Coast War Cry*, July 24, 1897

[3] From the forward of "A Study of the Homeless Man Problem in San Francisco" by W. S. Goodrich. Made for the Joint Committee on Single Men and Unemployment of the Council of Social and Health Agencies and the Community Chest of San Francisco, March–May 1924

Kang Solomon

Before there were homeless people sleeping in every other downtown doorway . . . before the Central Freeway was torn down . . . before Care Not Cash became a political catchword . . . before that, there was Kang Solomon.

It was the early 1980s. Kang was confused about picking up his General Assistance checks, so he lived in a cardboard box under the freeway next door to The Salvation Army's parking lot at 101 Valencia. From a distance you could see the accoutrements of his living space, a series of makeshift compartments that changed dramatically if the wind blew too hard or the rains came too suddenly. Blankets, couch, cooking utensils, a few clothes—all his personal belongings were carefully arranged on this plot of cement he called home.

Harry de Ruyter, Director of Social Services, along with his staff, Army officers and employees adopted Kang during this time—some with daily needs, some from a distance. Employees worried what could be done for him.

The answer, unfortunately, was very little.

The pattern of Kang's life was of his own making, according to de Ruyter. Kang had no wish to change. He lived on the street and that was his realm.

Mental health hospitals were scarce. There was little the Army could do to help except make him as comfortable as possible.

Kang was there in the mornings when the janitor, Bob Cox, checked the parking lot and there again when he left at night. For washing and doing his laundry, the building's downstairs bathroom near the switchboard was available. Sometimes we would meet him in the morning when his hair was matted and tangled and then again later after he had combed it neatly before venturing out. His downtown trips were on the first and fifteenth of the month, the day his checks came. On that day, he was dressed in a suit and tie, a sporty dresser strolling down Market Street. If he walked with the air of a royal, that's because he sincerely thought he was one. In the recesses of his mind, he was convinced he was King Solomon. And because he pronounced it and spelled it with a Southern accent, the Army helped get it officially spelled that way, so he would be willing to pick up his check.

He refused to accept it under any other name.

Salvation Army Addresses in the South of Market

The following addresses were gleaned from several sources—issues of *Pacific Coast War Cry*, *The Conqueror* and reverse telephone books. The date listed is where the reference was found and not necessarily the date the program opened.

1883 The Salvation Army arrived in San Francisco. Major Alfred Wells set up headquarters in his boarding house at 142½ Fourth Street
    Salvation Army at 12 Hampton Place
1885 White Wings at Third and Hunt Street and various locations in the South of Market offered food and shelter
    Pacific Coast Headquarters located at 1139 Market
1897 Wood yard, 564 Howard (Phone: MAin 1356)
    Relief Store, 826 Folsom
    Women's Shelter, 829 Folsom
    Relief Bureau, 863 Folsom
    Children's Home, 328 Bryant
    Rescue Home (Beulah) receiving office, 839 Folsom
1899 Corps on the Waterfront, 11th Street and 19th
1904 Industrial Home at 271-275 Natoma
1906 New Metropole at 147 Natoma
1916 Industrial Home at 852 –876 Harrison
1920 Western Territory Headquarters and Training School at 101 Valencia
1926 Red Shield Gymnasium, 95 McCoppin
1946 Adult Rehabilitation Center moved to 1500 Valencia
1950 Harbor Light Center moved to the old Southern Police Station, 360 Fourth Street, with 28 bed halfway house dormitory, 20 bed temporary shelter. Soup line.
1956 James House for Recovering Alcoholics opened in hotel next to Harbor Light with 30 single rooms for men.
1966 Southern Police Station remodeled to become "Bridgeway House," jail release program for recovering alcoholics from San Bruno Jail (22 beds)
1972 Harbor Light moved its social model Recovery Home for men (65 beds) to Gladding McBean Building, 1175 Harrison
    Social Services and Correctional Services at 176 and 178 Valencia
    Extensive outreach to prisoners at 850 Bryant
1974 Harbor Light Detox center opened for men
    Bridgeway Project moved to the Tenderloin
    Silvercrest Residence for Seniors built at 133 Shipley
1975 Women's Recovery Home component introduced at 1275 Harrison (14 beds for women and 51 beds for men)
1980 Harbor Light Detox opened for women
1982 Senior Meals and Nutrition Program operations at 360 Fourth
    Senior Meals site at the old Progress Building, 125 Valencia
1984 Centralized Social Services established at 407 Ninth Street
1989 Divisional Headquarters moved to 832 Folsom
1991 Harbor House Transitional Housing (formerly called Gateway) opened at 407 Ninth Street
1992 Senior Meals and Central Kitchen opened at 850 Harrison
    Asian American Corps opened at 832 Folsom
1998 Service Center opened at 520 Jessie between Sixth and Seventh Street, Howard and Mission
    Meals that Heal program operated from 850 Harrison
1999 Yerba Buena/Asian American Corps moved to 360 Fourth, meal site

EIGHT

Building Community

At the Heart of the Army

"It is at the local level that the world-wide heartbeat of The Salvation Army throbs because the officer here, of whatever rank or appointment, enjoys the sense of representing an international organization to a local community and to the individual he serves. Always he is at once part of this local pattern and of a global Army world."[1]

The First Chinese Salvation Army Corps In the World Was in San Francisco

San Francisco, port city to the Pacific, is the story of many peoples arriving in search of jobs and great expectations. Unfortunately, neither the jobs nor the expectations have always worked out. From the beginning, the Asian Pacific American community has been part of The Salvation Army family.

At the time of the Chinese Exclusion Act of 1882, anti-Chinese feeling ran high in San Francisco. The Salvation Army opened its work in the City in 1883 and by 1886, it had clearly focused on working with the Chinese population—men who had left their families in China and come to this country in search of work.

The Chinatown they lived in was full of opium dens, gambling, drinking, brothels and children sold into prostitution.

Into that melee marched The Salvation Army. On March 18, 1896, when the Chinatown Corps opened, Captain Fong Foo Sec sent a telegram to Commander Evangeline Booth to tell her the news: "The first Chinese Salvation Army Corps in the world is now an accomplished fact," he reported. "The hall is in a good location on Sacramento Street. Every seat is taken and a crowd stood on the sidewalk and looked through the glass doors, unable to get in."

In the early days, officers—mostly single young women—were often moved to Chinatown for short assignments, sometimes only three or four months. None spoke Chinese, although some tried. Nellie Keefe, Maud Sharp, May Jackson, May Thomas, Rebecca and Susanna Manhart, Margaret Pebbles and Pearl Stevens were among the early captains listed on the corps roster. The building at #2 Waverly Place was secured and dedicated on April 24, 1923.

Sue Tom remembers that Chinese called the Army the *"Boom Boom Wooey"* in the 1940s. That's the Chinese word describing the sound of the drum. As a young child she gravitated to the Army. Soldiers were often on the streets. So was she. With seven kids in her household, the streets of Chinatown were her playground. As a teenager, she joined the "Chinese Pioneers" (familiarly known as Chi-Pi) and participated in bazaars, fashion shows and choir rehearsals. Captain and Mrs. Walter Sleppy were in charge.

The building was remodeled and rededicated on March 25, 1958, under the leadership of Lieutenants Bill and Millie Lum. Lieutenants Check and Phyllis Yee arrived in 1959. The Army officially opened its new headquarters on November 12, 1972.

In addition to its regular programming, the Army offered residential housing for young Chinese women and then seniors in its building on Waverly Place. The corps was bursting at the seams. It needed more room.

Lt. Colonel William McHarg was a strong advocate for putting the Chinatown Corps into a new location and is remembered for securing grants and contracts to fund it.

Finally, after 50 years in the small, cramped quarters, the Army asked architect Thomas Hseih to design a new building. Six levels were designed to accommodate programs the corps saw in its future: a lower level community center with stage and kitchen, garage, group rooms and offices, a two-level gymnasium, space for housing, a TV studio and, above it all, the rooftop chapel. Leaded glass doors leading into the chapel show the continuum of service from Salvation Army founders, General and Mrs. William Booth, to the contemporary family. The panels were designed by Judy Raffael.

As new immigrants continue to arrive in Chinatown, young people have traditionally been a primary focus of programming that includes scouting, youth bands and singing groups. Summer day camp attracts more than 100 young people each year. Community basketball tournaments are held in the gym. A strong after-school tutorial program offers daily study opportunities to children from homes where English is not spoken. The program is intense and highly disciplined.

Sports, snacks and Friday night movies are on the agenda, but not until homework is done. Volunteer tutors are always needed.

The corps has a long history of encouraging interest in the arts, including the study of traditional Chinese instruments. During the era of in-house TV production, a generation of young people grew up with strong theatrical and technical production interests. These American Born Chinese—through highly-motivated Service Corps activities and musical tours in other countries—developed a strong Asian-American unit that eventually was ready to move out on its own in 1992. Captains Daniel and Susan Wun, Majors Robin and Tereasa Hu and Captains Thomas and Joy Mui have served as Chinatown corps officers.

The Oakland Chinatown Corps, also seeded from San Francisco Chinatown Corps, was dedicated on February 2, 1986.

Lt. Colonel Check Yee (R)

If he had stayed in China, it's likely Check Yee would have remained a journalist. He is an eloquent, sensitive writer.

If he and his wife Phyllis hadn't fled the Cultural Revolution in 1949 and moved to Canada, they might never have joined The Salvation Army.

If they hadn't been assigned to San Francisco's Chinatown, they might never have fostered an Asian-American ministry that also reaches back across the Pacific.

Lt. Colonel Check Yee—a member of The Salvation Army's Order of the Founder—was the second field officer in the country and the first in the Territory ever to be appointed to that rank. His record for longevity is legend. Before his retirement in 1994, he served thirty-five years in one assignment and one alone—San Francisco's Chinatown.

During that time, he and his wife moved Salvation Army services from an outgrown building at #2 Waverly Place into a large, multi-service facility at 1450 Powell. In the basement they hosted the American Cook's Training School (funded by the Mayor's Office), which prepared over 1,400 immigrants for jobs as professional chefs in San Francisco. Local restaurateur Johnny Kan made major donations of commercial equipment for the project. (Sadly, after many years, the American Cooks' Training School was closed because of health department restrictions. The space is now used for community meetings and educational programs.)

In the Army's rooftop chapel—surrounded by clerestory windows depicting the creation of the universe—the Yees presided over 500 weddings, 240 baby dedications and the enrollment of 700 senior soldiers. Across the hall the corps produced over nine hundred half-hour segments of *High Places,* a weekly KTSF-TV show for the Cantonese-speaking community in San Francisco.

It was the 18-year TV program that offered a ministry far beyond what the congregation could ever have imagined. Following the Mexico City earthquake of 1985, the corps went to the airwaves to raise $161,000 for relief aid. Following the 1987 earthquake and the great floods in China's Hunan Province in 1989, San Francisco Chinese responded again to a Salvation Army telethon appeal, this time raising $700,000 to build a hospital, school, housing and continued assistance to their kinsmen on the Chinese mainland.

Including special assignments after his wife's death and after his own retirement, Yee returned to China 10 times. His book, "Good Morning, China," researches the history of The Salvation Army in China.[2]

Asian-Americans, The Next Generation Is Electronically Amplified

ABC. That's shorthand for American Born Chinese.

Originally, all members of this breakaway corps were ABC, even though their inclusive title invited other Asian Americans to belong. "After all, we all eat from the same rice bowl," says Major Keilah Toy, daughter of Colonel Yee. She and husband Major Ron Toy have led the new corps from its earliest days.

The congregation has expanded. But it's not necessarily ethnicity that draws them together. It's music—and a style of worship with a distinctly contemporary, electronically amplified beat. Services attract a multicultural mix of young adults and families. Performing arts are a regular part of the agenda.

In The Salvation Army's world, this is a pretty hip place to be . . . !

Primarily, it's a congregation of working adults and professional people, families and a new generation of children. Group youth activities are stressed, both for children within the corps and for community outreach. A highly structured athletic program includes a regular basketball schedule with high schools in the Richmond and Sunset districts. Family events are first priority, including potluck wok suppers for any occasion from Sunday services to watch night events for New Year's Eve. Caroling is a Christmas Eve tradition. As many as fifty carolers visit seniors, disabled and AIDS patients at Children's Hospital, and then return to private homes with hot cider and refreshments.

Members of the technically savvy congregation have developed their own web site—www.asianamerican.org. After several years during which it met at the 832 Folsom headquarters building, the corps is now located at 360 Fourth Street and hosts a senior meals program for over 60 people daily.

Ejercito de Salvacion

In the early years, it appeared that officers took turns at a number of local corps. Many of the names at Coprs #4 are on the rosters of other units—Manhart, Goldthwaite, Ferguson, Smith, Polgreen, Crombie, Longdon, Malmberg.

In 1967, Commissioner William Parkins dedicated the new building at 1156 Valencia. That was the year Brigadier Carl Andreasen was in charge. After him came Captains Ken and Jolene Hodder, Majors Sal and Ida Gomez, and Major Martha Galbraith.

When Captains Larry and Nila Fankhauser arrived, it was still an Anglo corps. By the time they left five years later, it was well on its way to becoming fully Hispanic.

The building was sitting in the middle of the Mission District. It was only natural to reach out to a ready-made Hispanic constituency. First came a concentration on activities for women. One after another, neighborhood women joined and invited their friends. Following them, their children came to summer camp and after-school activities. Ultimately, the dads began to realize something really good was happening there as well.

Auxiliary Captains Pedro and Maria Yepez came to direct the Hispanic ministry as a wave of Hispanic corps opened in the Territory. Major Leslie Brooks followed the Fankhausers and when the Yepez's returned for four years as corps officers, Mission Corps #4 was a full fledged Hispanic corps. Majors Hector and Gerde Ramos and their family came in 1992 and were followed by Lieutenants Mario and Maria Ruiz who continue a strong community outreach.

Remembering Oceanview

In 1959, Lt. Marilyn Bawden's first assignment was 144 Broad Street. Like many small units, the Oceanview corps was financially strapped. The story she tells about keeping the doors open is typical.

In those days, Army officers collected money in taverns to support their work. Two nights a week she made the rounds of bars from the Richmond District to the Mission all the way out to the Cow Palace. She wasn't even 21 years old herself and had never been in a bar socially, but on a good night people would give her up to $200 to thank the Army for the youth programs she developed for neighborhood children.

Donations were a necessity. Her salary was $27.50 a week, and the rule was "pay the Army's bills before you pay yourself." Each Monday, she put $5 in a food jar. Each week she wondered how far the money would stretch.

The corps had opened in 1953. Lt. Roland Jones made a big splash introducing a black Santa Claus. Lts. Vickie Shiroma, Wes and Ruth Sundin and Captains David and Deanna Sholin were among the corps officers providing a primarily after-school program to children of working parents. In 1977 the Army sold the property to Faith Bible Church of San Francisco. In 1989, in the same neighborhood, it purchased the Halsted N. Gray Carew & English mortuary at 4000 Nineteenth Avenue for use as a corps/community center.

Korean Outreach

Although The Salvation Army had begun work in Korea as early as 1908, it was 1989 before a Korean contingent officially opened in San Francisco. It was a small band, meeting in the early days at a motel across from Divisional Headquarters at Valencia and McCoppin. Early officers here included Major Hwan Kwon Lee, Captains Man-Hee and Stephanie Chang, and Lts. Justin and Juhie Park.

Eventually, the congregation moved to a landmark-status cottage on the Great Highway and then to the building on Nineteenth Avenue, which before that was the Citadel Corps and before that the mortuary.

This is a close-knit unit, with Korean-Americans coming to San Francisco from as far as Richmond, California, to attend early morning services and participate in special projects. Over the years, they've become well known for going as a group to serve lunch at headquarters or dinner at the shelter on Eddy Street. On holidays they deliver meals to seniors and shut-ins.

One of the most successful outreaches during the tenure of Lieutenants Fred and Chris Kim was multicultural. Biennial Christmas concerts in Herbst Auditorium fostered relationships with Korean and other churches throughout the Bay Area, stressing the importance of recognizing and valuing the strengths of cultural diversity. Through summertime projects, soldiers established ties with Africa, starting with their first trip to Rwanda following the 1994 genocidal wars—a reminder of the terrible rifts many of the Koreans had experienced in the country of their birth. When the Kims were moved to start a unit in Santa Clara, they were followed by Captains Peter and Ok Kim.

Lighthouse Corps

One of the most important things about the Lighthouse Chapel at 445 Ninth Street is that it has two entrances. One opens onto the shaded Harbor Light/Harbor House courtyard where alcoholics in recovery play basketball and enjoy Fourth of July picnics and the good smells of breakfast, lunch and dinner emanating from the Sophie Hoffman/ Paul Handlery kitchen.

A second door to the chapel opens from the street and seems especially symbolic. This is an open door for the outside world to come join in the recovery efforts of the program going on inside. It represents the importance Harbor Light places on having a safe place for friends, relatives and program graduates to see it not just as a one-time residential program, but a healthy, ongoing recovery community.

Men and women on the program are not required to attend services, but many do. Coming to an understanding of a higher power, turning their lives over to God, as they understand him, is a major part of recovery.

The corps was officially opened September 30, 1992, under Majors Larry and Vickie Shiroma. Captain Lawry Smith was appointed first full time officer in charge.

From Speakeasy to Homeless Shelter
To Community Center

Sometimes it's the feeling of a building that you remember most. Pat Eberling—Divisional Social Services Consultant for twenty years—distinctly remembers walking through the 240 Turk Street building and being struck by the gaudy red flocked wallpaper on the walls. Bullet holes in the basement

Tenderloin Housing & Community Center proposed for a neighborhood that is home to 29,000 people, 3000 of whom are children. More than 57 percent of the residents live below the poverty line. The area has the highest rate of drugs and prostitution in San Francisco.

walls and dark, narrow hallways echoed with sounds of the speakeasy said to have been there in the past. Major Patrick Granat's first memories include a restaurant where the chapel and dining room now are, and a bakery for making pastries.

Ernie Marx, member of the Harbor Light Council and a realtor by profession, thought how expensive it would be to renovate. Perhaps only 60 percent of the building was habitable. Still, the Army was eager to develop a halfway house in the Tenderloin in the early 1970s. It leased the building for five years with the option to purchase. Originally, the program was under the wing of Harbor Light. By the time it became a corps in 1989, there were affordable studio apartments for people in recovery who needed a bridge back into normal living. The building was old. It wasn't the Ritz, far from it. But Bridgeway served as an inexpensive roof over the heads of many people at a crucial time in their lives.

In 1984, the Army made a decision to renovate the unused back of the building, that part which had probably led to the speakeasy. Ironically, the narrow alleyway through which bootleggers may have walked became a waiting area with benches where homeless men could wait for the Eddy Street late-night shelter to open. The door which opened during Prohibition for people looking for a drink now opened for men looking for shelter. A shower, clean pajamas, food, a clean bed and access to medical and social services were inside.

When the Army first made plans to move into the neighborhood in 1974, there was community concern. "Not In My BackYard" (NIMBY) sentiment was strong from those who thought Skid Row alcoholics would cross Market Street to add to the already drug-ridden streets.

But as the neighborhood has changed during the

Army's tenure, so have services. Under the leadership of Patrick and Kitty Granat when they were Envoys, Majors Neil and Kathie Timpson, Captains Barry and Serona Kaurasi and Lt. Roger McCort, the emphasis has now changed to youth.

Today, nearly 3,000 young people under 18 live within the immediate neighborhood, making a strong case for the state-of-the-art community center with swimming pool and full court gymnasium the Army plans for the site. It will be built in the same spirit as the community center the Army built with Joan Kroc funds in San Diego.

Working closely with San Francisco City and County representatives to determine greatest need, the Army has planned to provide supportive housing on separate floors to three basic groups—27 transitional units for "aged-out" foster care youth, 43 transitional units for adults in recovery and 40 permanent units for adults in recovery. (When they're no longer eligible for foster care, each year as many as 120 to 150 teenagers are suddenly on their own. National studies show that approximately 40% to 60% end up homeless within a year.)

Construction for the new building, scheduled for 2006, will tear down a structure long past its prime. A 1913–1915 Sanborn insurance map shows that the post-earthquake building at 240–242 Turk Street was called the Hotel Van Dorn and stood alone on the block except for the Bernard Apartments on Leavenworth and a cluster of lodgings at the corner of Leavenworth and Eddy. In the same series of maps for 1948–1951, the building was called the Hotel Dwaine and the block was tightly packed with buildings which, according to cops who worked the beat, were rumored to be connected by underground tunnels.

The same cops who know every inch of the street and alleys in between also know The Salvation Army. A visit inside the old building at noon captures in a minute what it does best.

Seniors from dilapidated single room occupancy hotels are in the dining room. They arrive early in the morning to save a seat, meet their friends, read the newspaper. They'll be back Friday to stand in line around the block for food distribution. Some will stop with the intake worker to ask for help. Stacked on one side of the room are meal delivery containers volunteers have taken to disabled seniors unable to leave their apartments.

As soon as the seniors are gone, the next meal shift arrives. It's a riotous group of neighborhood kids who sing happy birthday to a visiting board member and then sit amazingly still until the food arrives. When it's time for fun, chaos reigns. When it's time to listen, they listen.

What's it like to spend after school hours and summer vacation with The Salvation Army? Walk with these kids down the long, narrow hallway where the walls once were covered with red flocked wallpaper leading to the speakeasy. Remember how bare they seemed when the rooms were a late-night shelter for homeless men. Today, with kids clamoring in every direction, the change is awesome. The walls are covered with hand-painted flowers, trees, balloons, birds and animals in vibrant colors. In one room kids have painted flags representing their heritage—Mexico, Indonesia, Samoa, Ireland, Bosnia, Cambodia, Laos, Eritrea, Bangladesh, Vietnam, India, Ukraine, China, Korea, Germany, the Philippines. Youth Director Jennifer Arens is quick to point out the respect with which staff treats each culture. Some are the children of immigrants. Some, even though they've lived here for years, have never seen the Pacific Ocean just a mile and a half away.

The streets are congested. Walking through them is like making your way through a kind of gauntlet, with trouble—real or imagined—on either side. Definitely not for the faint of heart. With its drug traffic and loiterers, homeless and panhandlers, the Tenderloin seems a hopeless place.

Exactly the kind of neighborhood where The Salvation Army ought to be.

From One Generation to the Next

Usually, when the charismatic leader of a new movement dies, the energy and passion of the vision also withers and dies.

That hasn't happened with The Salvation Army.

From one generation to the next, William Booth's vision has held fast. Methods may have changed, program emphasis has changed, administration has changed, even the uniforms have changed. There are no more bonnets and no one sings "Hallelujah, I'm a Bum!"

Instead Salvation Army officers are known as no-nonsense property managers and good stewards of community resources. In a disaster they can mobilize on a dime. And always, always, the best of them know it's the personal connection that counts most.

These are quiet soldiers—three-dimensional beings, not *Guys and Dolls* stereotypes. They cry with people in trouble and share joy with them when things get better.

The images in this book are from San Francisco. In many ways, they're typical of The Salvation Army everywhere.

Notes

[1] Pilgrim's Progress: 20th Century, The Story of Salvation Army Officership by Senior Captain Don Pitt, Published by The Salvation Army National Research Bureau, New York City, 1950.

[2] References: "Galaxy of Glory," San Francisco Chinatown Corps; "For My Kinsmen's Sake" and "Good Morning, China" by Colonel Check Yee; "Marching to Glory, The History of The Salvation Army in the United States, 1880-1992," by Edward H. McKinley, 1995.

NINE

Who, What, When and Where

The Salvation Army "is the most effective organization in the United States. No one even comes close to it with respect to clarity of mission, ability to innovate, measurable results, dedication, and putting money to maximum use."[1]

Peter Drucker said it and you'd better believe The Salvation Army isn't likely to stop quoting him anytime soon!

When a management guru like Drucker is willing to chisel his approval of the Army in stone, people listen. Community leaders may wring their hands occasionally to see how carefully the Army moves in some directions, but when it comes to the bottom line, they know this is an organization that can squeeze money from a proverbial turnip at the same time as it produces intense loyalty from its fans and, without fanfare, helps a prodigious lot of people.

What's the secret? In partnership with Ben Brown from *USA Today*, former National Commander Robert A. Watson borrowed Drucker's words for a title and expounded on the thesis from his very personal point of view.

The book is "The Most Effective Organization in the U.S.: Leadership Secrets of The Salvation Army." It's a "how-to" essay combining smart business practices with a healthy dose of idealism and focusing on that which inspires people to work for more than the promise of a paycheck.

Watson makes it clear. The Army offers no stock options and no quarterly sales goals. All the traditional corporate incentives are out the window. What makes the Army work in his judgment is its humanity. At its best, it gives people a chance to practice small acts of kindness, the intangible rewards of which are huge incentives to want to do it again.

Watson stresses universal values and human aspirations. People want to feel good about the world. They want to feel as if they're participating.

They want to feel as if they're making a difference.

The Salvation Army offers a conduit for them to act on their best instincts.

A newspaper columnist helps because she remembers a pair of shiny Mary Janes given to her as a child on a Christmas shopping spree at the old Emporium on Market Street. A family donates in honor of a relative who stumbled into a Harbor Light program and finally walked out with enough dignity to start over again. A businesswoman remembers her mother's participation in Army activities in the Midwest. Two people meet at a fundraising dinner and start talking about other causes they support. I'll give to yours if you give to mine. It's a common bond.

Salvation Army advisory boards have no fiduciary responsibility. They're not corporate governing bodies. They're executives and community leaders who come to the table with advice based on corporate and civic experience. Yes, they generally come from a strong spiritual foundation, but not always Christian. Traditionally, Salvation Army boards throughout the country have often had a Jewish contingent.

In the last 25 years, board members have come from increasingly varied backgrounds. One, for instance, laughingly refers to himself as being "to the right of Genghis Khan." Another is a "moderate progressive" Democratic Party leader.

Instead of their names, think of them for who they are—corporate managers, educators, accountants, car dealers, hotel executives, a newspaper executive, an NFL referee, a florist, a wholesale grocer, realtors, a union leader, generals and rear admirals, high tech systems analysts, medical doctors, nurses and a dentist, a fudge brownie specialist, former mayors, an industrial designer, advertising and public relations professionals, a former presidential speechwriter and long-range strategic planner, a TV station owner, marketing consultants, judges, venture capitalists, small business owners, stock brokers, financial planners, a City Attorney and a City Supervisor . . . people who care about the community.

"There are Four Salvation Armies in this Country, Not Just One . . ."

In addition to local advisory organizations, a National Advisory Board advises the Army's Commissioner's Conference and meets three times a year.

Territorial Commanders of The Salvation Army's four independent and very diverse United States territories make up the Commissioners' Conference. In effect, said San Franciscan Bill Waste who served at both the local and national level for over thirty years, "there are four Salvation Armies in this country, not just one. The National Commander serves as the fifth commissioner, an equal among equals." Waste's service was mostly on business committees dealing with personnel, fiscal responsibility, investment programs and the national portfolio. He and Jack Barnard of Bechtel worked on the leveraging of purchasing power, including fleet management practices.

The job of the conference is to guide the Army's overall direction. A majority vote is required on all issues.

Veteran San Francisco board members Bob Alspaugh, Mary Theroux and Rob Pace currently serve on the National Board.

Note

[1] Peter Drucker quoted in "Peter Drucker's Picks" by Robert Lenzner and Ashlea Ebeling, *Forbes Magazinee*, August 11, 1997

Chronological Highlights During the Lifetimes of William and Catherine Booth

1829 Catherine Mumford born on January 17 in Ashbourne, Derbyshire, England. William Booth born on April 10 in Nottingham, England.

1837 Victoria became Queen of England.

1855 Catherine and William married on June 16 at Stockwell New Chapel, London.

1858 William Booth ordained a Methodist minister, May 27.

1860 Catherine's first public speech, May 27, Whit Sunday.

1861 Charles Dickens wrote Great Expectations. Beginning of the American Civil War.

1865 Booth founded the Christian Mission in East London.

1878 Name of the Christian Mission changed to The Salvation Army.

1879 Army work begun in Scotland and Channel Islands.

1880 Army work begun in America, Ireland and Australia.

1881–1882 Army work begun in France, Canada, India, Switzerland and Sweden.

1883 Army work begun in San Francisco, California, on July 22. Also in Sri Lanka, South Africa, New Zealand, the Isle of Man and Pakistan. The first prison-gate home opened in Australia.

1885 Salvation Army instrumental in influencing bill in London to raise the age of consent to 16, a blow to international traffic in underage prostitutes.

1886 Work begun in Newfoundland and Germany. First slum corps opened in Walworth, London.

1887 Work begun in Italy, Denmark, Netherlands and Jamaica.

1888 First food depot opened in Limehouse, London, and young people's work organized throughout England. Work begun in Norway.

1889 Army work begun in Belgium and Finland.

1890 Catherine Booth "promoted to Glory" on October 4.
"In Darkest England and the Way Out" published.

1891 Industrial Colony established in Hadleigh, Essex. Work begun in Zimbabwe and Zululand.

1894 William Booth's visit to San Francisco. Army work begun in British Guiana, Iceland, Japan and Gibraltar.

1897 First Salvation Army hospital founded at Nagercoil, India.

1898 Fort Romie Farm Colony opened on January 6 near Soledad, California Additional farm colonies opened: Fort Amity in Colorado, Fort Herrick in Ohio. Work begun in Barbados and Alaska.

1904 Army work begun in Trinidad, St. Lucia, Grenada, Antigua, St. Vincent, Panama and St. Kitts.

1906 San Francisco earthquake and fire (April 18-20) Extensive Salvation Army relief services in Oakland and San Francisco.

1908 Anti-suicide Bureau and Home League established. Booth received by Kings of Denmark and Norway, Queen of Sweden and Emperor of Japan and earned an honorary degree from Oxford.

1910 Leprosy work begun in Java, and work begun in Chile, Peru, Paraguay and Sumatra.

1912 William Booth promoted to Glory on October 20 in Hadley Wood, England.

Compiled from The Salvation Army Yearbook and other sources

Among Those Who Have Served On the San Francisco Advisory Board

Chairs (in alphabetical order): Robert Alspaugh, William Banker, Jack Barnard, Herb Cunningham, James Deitz, James Dinwiddie, Art Fairchild, Paul Handlery, Michael Hardeman, Sophie Hoffman, Arch Monson, Henry Morris, Richard Osgood, Jack Podesta, Robert Starzel, Mary Theroux, William T. Waste.

Alphabetically again, some of the stalwarts from recent years include: Judge Carl Allen, Joe Allen, Robert Appleby, Judge Albert Axelrod, Robert Alspaugh, Pamela Baldwin, William Banker, Jack Barnard, Ernst Bauer, Robert Begley, Steve Bell, Brooks Berlin, Rear Admiral John Bittoff, Phil Bowhay, Justice A.F. Bray, E.D. Bronson, Marie Brooks, Annette Campbell-White, Marvin Cardoza, Harvey Carter, Mrs. Allen Chapman, Frank Chong, Mayor George Christopher, Peter Chung, Richard Clark, Douglas Cornford, John Covert, Fred Parr Cox, Hilary H. Crawford, J. D. Crowley, Herb Cunningham, Earle Dahlem, Owen Davis, Bert Decker, Dr. James Deitz, Alvin Derre, Richard Dinner, James Dinwiddie, Hayley Ditzler, Prudence Dorn, Jane Eckels, Victor Eggerding, Gordon Elliott, Art Fairchild, Adrian F. Falk, Senator Dianne Feinstein, Jesse Feldman, Tiffany Franchetti, Herbert Funk, Al Gee, Mrs. J.C. Geiger, George B. Gillin, Mike Goefft, Mrs. Henry F. Grady, General David Grange, Pat Gray, E. G. Grayson, Susan Green, W. Lawrence Greer, Paul Handlery, George Hansen, Ken Hansen, Michael Hardeman, Alvin Hayman, Kristin Heuter, Doug Hiemstra, Sophie Hoffman, Fifi Holbrook, Nikki Howell-Chicotel, Dr. Fred Hudson, Mayor Frank Jordan, Warren Kelleher, Judge Joseph Kennedy, Herman Kersken, Karl Kimbrough, Graham Kislingbury, Bonnie Kreiling, Robert Lacovic, John Lambert, Richard and Robert Lautze, Wade Leach, Judge Gerald Levin, William Lee, Vernon Libby, John Livingston, O.G. Linde, Kenneth Louth, Meta McDowell, Al Maggio, Lt. General Glynn Mallory, George Mardikian, Dick Martin, Fred Martin, Edward Mazyck, Chauncey Medberry, Martin Michael, Gray Minor, Arch Monson, Jr., Lt. General James E. Moore, Jr., Joe Moore, Robert W. Moore, Henry Morris, Evelyn Mulpeters, James Mulpeters, Gerald Napier, Wendy Nelder, Mrs. Berkley Neustadt, Karen Osborne, Cathy Osgood, Richard Osgood, Robert Pace, John W. Pettit, Alfred Pfeiffer, Jack Podesta, Fred Postel, Eddie Powell, Jeffrey Qvale, Professor Marshall Raines, Peter Ratto, Judge Timothy Reardon, Former City Attorney Louise Renne, Donald Reid, Richard M. Robbins, Judge Henry Rolph, Jeanine Scancarelli, Daniel Schmitt, William Selover, Lou Simon, Eric Stanford, Alden Stanton, Robert Starzel, Leonard Stec, William Steele, Joe Sweeney, Ben Swig, Steve Swig, Peter Tamaras, Mary Theroux, Joseph Thompson, Burl Toler, Admiral Robert Toney, Dr. Wellman Tsang, J. Francis Ward, William E. Waste, William T. Waste, Erik Weir, Eddie Whitehead, Robert Wilhelm, Bob Werbe, Kenneth Whitman, Robert Whitman, William Wilkinson, William Wittick, Lt. General Fred E. Woerner, Mrs. Roy Wolcott, Pablo Wong, William Wong, Claes Wyckoff, David Yoder and Norma Young.

Councils and Special Committees

Adult Rehabilitation Center—Lou Batmale, Peter Bogardus, Bud Burtis, Richard Clark, Evie Dawe, Wesley Dawe, Carol Ede, Lee England, Charles F. Gunther, Grant Hellar, Tim Hill, Charles Johnson, Woodward Kingman, Myra Kopp, Patrick McNerney, Judge Henry Rolph, J. Curtiss Taylor, Paul Vogelheim

Annual Luncheon Planning Committee—Robert Alspaugh, Ken Hansen, Mike Hardeman, Sophie Hoffman, Arch Monson, Henry Morris, Wendy Nelder, Eddie Powell, William Wilkinson

Emergency Disaster—Admiral John Bitoff, Sophie Hoffman, Wendy Nelder

Pinehurst Auxiliary—Pam Abbis, Simone Alliguie, Peggy Anderson, April Andrews, Tiny Boyd, Lavonne Bullis, Carol Canellos, Alice Carouba, Peggy Cavanaugh, Grayce Ceshi, Juanita Cunningham, Yvonne Daubin, Diana Davis-Lopez, Felicie Del Bonta, Francis Dickie, Becky Donnelly, Prudence Dorn, Carol Ede, Marlo Erickson, Peggy Friedman, Dolly Garcia, Lita Goring, Carol Harlan, Rita Heiser, Liz Helman, Anita Hilmoe, Dorothy Jack, Margot Jacobs, Jeannette Johnson, Joy Johnson Kay Klauber, Althea Klingel, Jane Lefferdink, Irene Loupy, Carolyn Marcus, Adrienne McKelvie, Pat Mertens, Ollie Murray, Janice Pinkston, Signe Rianda, Claudia Richardson, Dorothy Ritter, Linda Roberts, Buzz Rolph, Samara St. Denny, Audrey Schmitt, Jeanette Shahan, Lois Storz, Suzanne Strauss, Bobbe Vargas

Chinatown Corps—Helene Ah-Tye, Wayne Chan, Lily Chin, Mary Yee De Brunner, Jane Fong, Al Gee, Barbara Ann Lee, Della Lee, Howard Lee, Evan Leong, Elaine Leung, William Lee, Frank Lee, Richard and Robert Lautze, Candy So, Beatrice Wong, Helen Wong, Shelton Wong

Gateway/Harbor House—Robert Alspaugh, Steve Bell, Steve Blackburn, Annette Campbell-White, Susan Green, Janet Reilly, Alfred Pfeiffer. Karen Osborne, William Wilkinson

Harbor Light—Judge Albert Axelrod, Robert Batchelor, Joe Betz, Jack Block, Harvey Carter, Donna Casey, Bruce Christensen, Dr. Trish Connolly, Mary Yee De Brunner, James Dinwiddie, Ed Flowers, Jr., Al Gee, Al Graf, Noah Griffin, Paul Handlery, Paul Harrell, Sophie Hoffman, Dr. Fred Hudson, Grant Hundley, Dr. Marissa Israel, Mayor Frank Jordan, Lona Jupiter, Harold Kilian, Karl Kimbrough, Graham Kislingbury, Terry Lowry, Roy Maloney, Ernie Marx, Michael Milstein, Robert W. Moore, John Pettit, Lawrence Pong, Peter Ratto, Al Stanton, Len Stec, Ben Swig, Walter Tingley, Edward Wright, David Yoder

Long Range Planning—Clare Gordon, Karen Osborne, Dan Schmitt, William Selover, William Wilkinson

Mission Corps—Ishmael Burch, Bonnie Kreiling, Patrick Mora, Enicia Montalvo, Gilberto Nunez, Patrick Preminger, Theresa Reid, Arthur Rugama, Jon Shepherd, Pablo Wong, Norma Young

National Advisory Board—Robert Alspaugh, Jack Barnard, Robert Pace, William Waste

Nominating—Henry Morris, William Selover, Robert Starzel

Public Relations—Jane Eckels, Mike Goefft, Clare Gordon, Fifi Holbrook, Graham Kislingbury, Meta McDowell, Martin Michael, Marshall Raines

Senior Meals and Activities—Pamela Baldwin, Jack Barnard, Chester Bowles, Prudence Dorn, Nikki Howell-Chicotel, Wesley Dawe, Paul Handlery, George Hansen, Sophie Hoffman, Sergeant Frank Hutchens, John Kimbro, Karl Kimbrough, Fred LaCosse, Bill Lee, Bonnie McRobbie, Arch Monson, Martin Michael, Frank Ponder, Marshall Raines, Burl Toler, Jack Wolf.

San Francisco Women's Auxiliary—Rose Astor, Bessie Baughn, Jane Char, Jean Chu, Mary Cordini, Evelyn Dawe, Prudence Dorn, Helen Fong, Irene Gee, Pat Gray, Betty Hersch, Sophie Hoffman, Bertha Jones, Elsa Jones, Dorothy Kislingbury, Harryette Koch, Barbara Ann Lee, Mary Jane Lee, Della Lee, Ida Lee, Jane Lefferdink, Blanca Lemus, Virginia McCrostie, Gertrude Marx, Lois Molkenbuhr, Joan Morris, Evelyn Mulpeters, "Buzz" Rolph, Connie Skinner, Joann Stanton, Mignon Tau, Sue Tom, Laura Waste, Ellen Wing, Bea Wong, Helen Wong, Violet Wong, Hazel Yuen.

Yerba Buena Center—Evelyn Dexter, Anita Hill, Lionel and Kathryn Spencer

Northern California and Nevada Division

Geographic Area:

1920 Northern California and State of Nevada (except the City of San Francisco).

1924 Northern California and State of Nevada (except Bay Area and coast towns to San Luis Obispo.)

1948 Northern California and State of Nevada (except San Francisco.)

1957 Northern California (as far south as Bakersfield and San Luis Obispo)
State of Nevada (except Las Vegas).

Divisional Headquarters Locations:

1920	417 Market Street, San Francisco, CA
1921	115 Valencia Street, San Francisco, CA
May 1921	706½ "K" Street, Sacramento, CA
1922	912 8th Street, Sacramento, CA
1929	429 "J" Street, Sacramento, CA
1929	1120 5th Street, Sacramento, CA
1932	1160 Valencia Street, San Francisco, CA
1933	533 9th Street, Oakland, CA
1940	506 15th Street, Suite 404, Oakland, CA
1944	1540 San Pablo, Suite 904, Oakland, CA
1946	1145 Foothill Blvd., Oakland, CA
1956	1439 Alice Street, Oakland, CA
1961	60 Haight Street, San Francisco, CA
1976	101 Valencia Street, San Francisco, CA
1989	832 Folsom Street, San Francisco, CA

Divisional Leaders:

1920	Lt. Colonel Arthur Brewer
8/25/20	Brigadier Andrew Crawford
1921	Major Allison Coe
2/1/23	Brigadier John Hay
8/25/27	Staff Captain David Boyd
1/25/29	Brigadier Louis C. Bennett
1/25/31	Major Arthur D. Jackson
7/11/33	Major Ernest Higgins
1/23/36	Major Holland French
9/20/38	Lt. Colonel James C. Bell
7/30/42	Lt. Colonel Arthur D. Jackson
9/14/44	Brigadier Arthur Brewer
7/53	Lt. Colonel Reginald E. Martin
10/57	Lt. Colonel Ranson Gifford
1/64	Lt. Colonel William J. McHarg
	Lt. Colonel Inga McHarg, Director of Women's Services
9/70	Lt. Colonel Robert Angel, Divisional Commander
	Lt. Colonel Gertrude Angel, Director of Women's Services
9/73	Lt. Colonel Victor Newbould, Divisional Commander
	Lt. Colonel Ardys Newbould, Director of Women's Services
7/81	Lt. Colonel Ray Robinson, Divisional Commander
	Lt. Colonel Bunty Robinson, Director of Women's Services
8/86	Lt. Colonel Bruce Harvey, Divisional Commander
	Lt. Colonel Dorothy Harvey, Director of Women's Services

Golden State Division

Geographic area:
San Francisco coastal area inland to Bakersfield
- 34 corps
- 1 Harbor Light Center
- 8 social service departments
- 1 outpost
- 66 service units
- 11 institutions
- Headquarters at 832 Folsom Street

Divisional Leaders:
1995 Major Jerry Gaines, Divisional Commander
- Major Jeanine Gaines, Director of Women's Services

1997 Lt. Colonel Richard Love, Divisional Commander
- Lt. Colonel Bettie Love, Director of Women's Services, San Francisco City Administrator

2004 Major Joseph Posillico, Divisional Commander
- Major Shawn Posillico, Director of Women's Ministries, Candidates Secretary, Officer Care and Development Secretary

Questions People Usually Ask

Who are these soldiers?
Salvation Army soldiers are members of The Salvation Army church. Just like Presbyterians, Methodists, Zen Buddhists, Muslims or any other religious group, during the week they may be medical doctors, bus drivers, accountants, homemakers, musicians, marathon runners, graphic artists, city employees or even the director of the Rose Bowl parade. On Sundays they worship in a Salvation Army corps/community center which keeps its doors open during the week as a social service center.

Do they wear uniforms to church?
Yes, if they want. Uniforms are a way of showing unity for special occasions and volunteering for disaster services.

What's the church like?
Pretty much like a Methodist church. That's where the founder William Booth was preaching before he started out on his own.

Who's an officer?
This is a soldier who agrees with the Army's beliefs and is serious enough about its mission to go into the work full time. When people ask about joining the Army, they're invited to attend services over a period of time, become familiar with its tenets, its structure, its people. "Take your time," said William Booth to his future son-in-law Frederick St. George de Lautour Tucker when he wanted to join the Army. "First of all, discover everything about us for yourself."[1]

As in other denominations, candidates are sponsored by the local congregation to go through two years of training college to become officers (ordained ministers) of The Salvation Army church.

And an envoy?
A person of a certain age who has retired from other lines of work or life experiences and joined The Salvation Army as a second career.

What's the difference between a captain and a major?
A captain is any officer in the first 15 years of service. Note the uniform. It has plain scarlet epaulets. A major has served more than 15 years and adds Salvation Army crests to the scarlet epaulet.

How do you get to be a colonel?
The old fashioned way—you earn it. Colonel is a title usually associated with a specific assignment such as Divisional Commander. Uniforms add silver piping to the scarlet epaulets with a crest.

What about women?
Same thing. Women have always held equal rank in The Salvation Army. Thank Catherine for that.

What's a Commissioner?
The Salvation Army in the United States is divided into four territories—Western, Central, Southern and Eastern. Each is led by a commissioner who, as the territorial commander, has final authority over his or her area.

Is There a National Commander?
A fifth commissioner is the national commander who represents the Army nationwide. Commissioners meet as a group three times annually at the Commissioners' Conference held in cities around the United States.

And the General?
There is only one general. He or she has overall responsibility for the Army's work worldwide.

What if I don't recognize the rank?
When in doubt, choose the higher rank. In The Salvation Army, most people's first name is Major!

Where does the power lie?
It depends on the issue. As a quasi-military operation, the Army can move extremely fast in case of an emergency

with orders coming from the top. In everyday matters, local officers have a great deal of latitude to speak to the needs of a local community. Matters of overall basic Army policy are determined at the international level and that, of course, sometimes takes a while to arrive at consensus from leaders of the more than 100 countries in which the Army serves.

In the West, there's one corporate board of directors
The Salvation Army's Western Territory is a single corporation with a single Board of Directors acting on issues which affect Alaska, Arizona, Colorado, California, Guam, Hawaii, Idaho, the Marshall Islands, Montana, Nevada, New Mexico, Oregon, Utah and Washington.

Where is The Salvation Army's church?
In keeping with the Army's military parlance, its churches are called corps/community centers. A corps is an organizational unit (like Marine Corps or Corps of Engineers). It's a center of worship. And it's also a community center, meaning that the building is expected to do double duty as a social service provider.

Everything happens here
Salvationists say they try hard to practice what they preach. Nowhere is it more apparent than in the multi-service use of their buildings.

Like Puritans, Quakers, Shakers, Methodists, Presbyterians and Zen Buddhists, they basically see their places of worship without the accoutrements of high church. Buildings are usually plain and serviceable, without stained glass, without extra décor. That's because there's no telling how many other uses the room will get.

In San Francisco, for instance, USDA commodity food is distributed every Friday from corps lobbies in the Mission District, South of Market, Chinatown and Tenderloin. The Tenderloin Corps has been known to set up cots in its chapel to house fire victims.

A wide variety of people call the Army their church—people with a commitment to community service as well as people who've been through pain themselves and feel comfortable in a place where human frailties are recognized honestly, people who have succeeded just enough to know how hard it is for others to do it.

Note

[1] "No Discharge in This War" by General Frederick Coutts, 1974

International Mission Statement

The Salvation Army, an international movement, is an evangelical part of the universal Christian Church.

Its message is based on the Bible. Its ministry is motivated by love for God. Its mission is to preach the gospel of Jesus Christ and meet human needs in his name without discrimination.

For Further Information

**The Salvation Army
Golden State Division**
832 Folsom Street
San Francisco, CA 94107
(415) 553-3500
www.tsagoldenstate.org

**The Salvation Army Crestmont College
Elftman Memorial Library**
30840 Hawthorne Boulevard
Rancho Palos Verdes, CA 90275-5301
Misty Jesse, Director
(310) 544-6475
E-Mail: Misty_Jesse@usw.salvationarmy.org
www.crestmontcollege.edu/library

**The Salvation Army Museum of the West
Crestmont College**
30840 Hawthorne Boulevard
Rancho Palos Verdes, California 90274
Captain Linda Jackson, Director
Phone: (310) 541-6241
E-Mail: Linda_Jackson@usw.salvationarmy.org

**The Salvation Army National Headquarters Archives
and Research Center**
615 Slaters Lane, P.O. Box 269
Alexandria, VA 22313-0269
Susan Mitchem, Director
Phone: (703) 684 5529
E-mail archives@usn.salvationarmy.org

**International Heritage Centre
The Salvation Army**
P.O. Box 249, 101 Queen Victoria Street
London, EC4P 4EP, England
House 14, William Booth College
Denmark Hill, London SE5 8BQ
Major Christine Clement, Director/ Gordon Taylor, Archivist
Phone: 020 7737 3327
E-mail heritage@salvationarmy.org
www.salvationarmy.org/history

Books about The Salvation Army printed by Crest Books can be purchased through Salvation Army Supplies and Purchasing departments:
Atlanta, GA – (800) 786-7372
Des Plaines, IL – (847)-294-2012
West Nyack, NY – (888) 488-4882

Retrospective

Early Christmas dinner in San Francisco, where the first holiday meal was instigated by Captain Joe McFee.
Artwork courtesy of The Salvation Army Museum of the West

***The three-day fire** after San Francisco's 1906 earthquake was so intense the heat could be felt in Oakland, where The Salvation Army ran a relief camp for refugees from the city.*
Photo courtesy of the San Francisco Chronicle.

In the early days, *lonely wanterers knew The Salvation Army in cities throughout the country. Today's transients are younger and often have multiple diagnoses. Courtesy of The Salvation Army Museum of the West.*

*During the Depression, lines of people in every city came to Salvation Army soup kitchens.
Photo from The Salvation Army Museum of the West, city unknown.*

*Salvation Army units throughout the country welcome volunteers.
Early Christmas dinner in San Francisco.*

Volunteers are partners in all the Salvation Army's work.

Music is an integral part of The Salvation Army.

Index

101 Valencia, 74-76, 78, 81, 83, 11, 119

Abe, Tozo, 81
Adams, Greg, 2
Adult Rehabilitation Center, 52, 69-73
Advisory Board membership list (past and present), 134
Arens, Jen, 129
Asher, Joseph, 27
Auxiliaries, Councils and Committee membership lists (past and present) 135
Afshar, Mike, 118
AIDS, 67, 68, 84, 87, 108, 126
Alcoholism Treatment, 57-73
Alger, Horatio, 37
Almeida-Oetting, Luisa, 51
Allemann, Elsie, 37
Alspaugh, Robert, 132, 134, 135
American Red Cross, 36, 43, 52, 55
Angel, Gertrude, 136
Angel, Robert, 74, 83, 136
Argonaut, 22
Army of Stars, 112
Ash Barrel Jimmy, 57
Asian American Corps, 126
Avery, Kasson, 87
"Away With Rum, By Gum", 58
Ayala de Gavidia, Gloria, 49

Baker Place, 67
Bankhead, Tallulah, 1
Barbary Coast, 14, 21
Barnard, Jack, 132, 134,135
Barton, Bruce, 2
Baugh, Alice McAllister, 40
Baur, John, 74
Bawden, Marilyn, 80, 126-127
Bawden, Ron, 80
Bay City News, 48
Bay Street Jazz Band, 106
Beacon, The, 70
Bearchell, Catherine ("Sammy"), 46
Barry, Harold, 46
Beulah Park, 20, 34, 38
Bishop, Mary, 100
Bleu, Don, 88
Block, Jack, 66, 135
Booth, Ballington, 6

Booth, Bramwell, 3, 69
Booth, Catherine, 3, 6, 8, 9-11
Booth, Catherine ("La Marechale"), 11
Booth, Catherine Bramwell, 11
Booth, Evangeline, 6, 10, 34, 36, 39, 76, 107, 111
Booth-Hellberg, Lucy, 6
Booth, Herbert, 6
Booth Home, 76, 82
Booth, Marian, 6
Booth, William, 1-9, 12, 14, 44, 69, 93, 96, 101, 111, 125
Booth-Tucker , Emma, 6
Booth-Tucker, Frederick St. George de Latour, 138
Bourne, Alice, 106
Bowler, David and Jean, 120
Brannan, Sam, 14
Brewer, "Buzz", 84
Brooks, Leslie, 126
Brorsen, Bob, 96
Bridgeway, 67, 128
Broughton, Bruce, 13
Broughton, William, 13
Brown, Antoinette, 9
Brown, Willie, 92
Brugman, Dorothy Koerner, 45
Buchanan, Chris, 30, 99
Burrows, Eva, 11
Butler, Anna Alleman, 18, 36

Cab Horse Charter, 8
Caen, Herb, 1, 87, 90
Camp Redwood Glen, 88
Caring for Children, 51
Celebrity Bell Ringers
   Media personalities, sports figures, community leaders, 89-94
Central Kitchen, 118
Charioteers, 25
Chang, Man-Hee and Stephanie, 127
Chavez, Evelyn, 55
Chinatown Corps/Community Center, 124-26
Chinese Pioneers, 124
Chinese War Cry, 18, 24
"Christianity with its sleeves rolled up," 12
Christmas —
   First kettle, 105

   First dinner, 105
   First bell, 109
Christmas in the Barrio, 48
Church, George, 29
Cline, Grace Phillips, 30, 91
Cline, Therma, 76
Clitheroe, Cyril, 67
Collier, Richard, 12
Cornerstone, 76
Community Chest, 116
Council of Armenian-American Organizations of Northern California, 104
Cox, Margaret, 81
Crawford, Rheba, 2, 110, 111
Crestmont College, 11
Crittenton, Charles, 20
Cypress Lawn Cemetery, 4

Dahlberg, David, 55
D'Anger, Yvonne, 83
"Daughter of the Regiment," 77
Davey, Stan, 59, 67
Dawe, Evelyn, 76, 135
Day, William, 38
DeRuyter, Harry, 120, 122
Dear Abby, 82
Decatur, James, 104
Depression, The, 75-77
Detox, 64-65
Devil's Island, 84
Dexter, Evelyn, 118, 135
Dexter, Ray, 80
Devine, Amelia Kunkle, 90, 103, 110
Dewson, Ruth, 95
Dingman, Frances, 5, 56, 77, 85
Dinwiddie, James, 66, 134, 135
Divisional Leaders
   Northern California and Golden State. 136-37
Doctor, Lloyd, 112
Docter, Robert, 100, 112
Doda, Carol, 83
Domestic Partners, 68, 118
Doughnut Girls, 40-43
Doughnut recipes, 42
Dowdle, James, 96, 111
Drucker, Peter, 93
Dunmore, Claire, 100
Duplain, George, 47, 70

INDEX 149

EASE, 118
Early corps, 18-19
Eberhart, Pauline, 10, 83, 88
Eberling, Pat, 127
El Salvador earthquake, 49-51
Elaw, Zilpha, 9
Elizabeth II, Queen of England, 3
Emperor Norton, 20
Evangeline Residence, 76, 83
Evans, Willard, 78

"Falling off the wagon," 62
Farm colonies, 8
Fankhauser, Larry and Nila, 126
Farrar, W.W., 9
Feinstein, Dianne, 120, 134
Ferguson, Harold, 67
Florence Crittenton Homes, 20
Foley, Cindy, 97
Fong Foo Sec, 17, 124
Food For Friends, 99,119
Fort Romie, 8
Franciscans, 2, 4
Free speech, 20, 102
French, George, 35, 38-9
Friendship House, 68
Froderberg, Trish, 41
Froderberg, Wayne, 76, 80
Frost, Barry, 112-113
Fry, Charles William, 96

Gable, Clark, 37
Gabriel, Nick, 67, 144
Gabriel, Ruth, 67
Gaines, Jeanine, 137
Gaines, Jerry, 29, 137
Galbraith, Martha, 126
Gap, The, 121
Garcia, John, 51
Gay/Lesbian/Bisexual/
    Transgender Task Force, 67
Gifford, Adam, 19, 74
Gifford, Ranson, 83
Gillespie, David, 27
Ginn, Sam, 120
Goldthwaite, Emma, 39, 45
Gomez, Ida, 126
Gomez, Sal, 84, 126
Gonzalez, Joe, 67
Gowans, John, 11, 13, 112
Graham, Carrie, 67
Granat, Patrick, 128-9
Granat, Kitty, 129
Grant, Cary, 2
Guatemala earthquake, 47-48
"Guys and Dolls," 1, 2, 103, 109-111, 130.

Haight Street, 83
Hammond, Beatrice and William, 41
Ham and Eggs Fire, 37
Handlery, Paul, 127, 134, 135
Hansen, Gladys, 32
Harbor House (formerly Gateway)
    119-20
Harbor Light, 64-68, 118
Harvey, Bruce, 29, 53, 99, 120, 136
Harvey, Dorothy, 120, 136
Hayburn, Robert, 27
Hayman Construction, 120
Hearst, William Randolph, 38-39, 82
"Heartbeat Theater, 112
Helms, Craig, 13
Helms, Gordon, 13, 53
Henderson, Ernie, 31
Herb Caen All Weather All Brass Band, 89
Hill, Gerald and Suzanne, 84
Hill, Grace Livingston, 43
Hirahara, Takamaru, 81
Hodder, Ken and Jolene, 126
Hoffman, Sophie, 88, 94, 99, 127, 134, 135
Holbrook, Fifi, 90, 134, 135
Home League, 84
Home Service Appeal, 2, 44
Homeless, 36, 69, 107, 116, 117, 120-122, 128
Hope, Bob, 109
Hostetler, Larry, 120
Hseih, Thomas, 125
Hu, Robin and Tereasa, 120, 125
Hudson, Brian, 67
Hunt Street, 66, 117
Hurricanes and other disasters, 52
Hutchins, Frank, 118, 135
Huxley. T.H., 9

Imai, Masahide, 81
"In Darkest England and the Way Out,"
    8, 9
India, 14
Indonesia, 14
Industrial Home and Yard, 69, 70
International Soundex Reunion, 82
Iremonger, Ken, 31,118
Ives, Charles, 13

Jackson, Helen, 46-47
James House, 66
Japanese Children's Home, 81
Japanese Division, 80-81
Joe the Turk, 20
Jones, Jerry, 112

Jones, Marlon, 96
Jones, Roland, 127
Jordan, Frank, 134, 135

K101, 88
KCBS, 93
KFRC, 93, 94
KGO-Radio, 48, 94
KGO-TV, ABC-7, 49, 91
KNBR, 91, 95
KPIX-TV, 84, 91
KRON-TV, 48, 93, 99, 119
KTSF –TV, 125
KTVU-TV, 93
Kaiser Medical Center, 68
Kan, Johnny, 68
Kang Solomon, 183
Kaurasi, Barry and Serona, 129
Kim, Fred, 127
Kim, Chris, 31, 127
Kim, Peter and Ok, 127
Kislingbury, Dorothy, 88, 135
Kislingbury, Graham, 95, 134, 135
Knights of Pythias, 74
Kobayashi, Masasuke, 4, 80
Koerner, Henry, 43-45, 76, 81, 95
Koerner, Marie, 4, 44-45, 95
Korean Corps, 127
Kroc, Joan, 129

Lange, Dorothea, cover, 60-61
Lansing, Arlee, 45
Lantz, Elsie, 31, 84
Lantz, Gene, 31,118
Larsen, Harry, 78, 111
Larsen, Dot, 80
Larsson, John, 13, 112
Laurel Hills Cemetery, 36
League of Mercy, 108
Lee, Barbara Ann, 88, 135
Lee, Jarena, 9
Lee, Mary Jane, 30, 88, 135
Lee, Patsy, 17
Lee, Hwan Kwon, 127
Leslie, Evangeline (Moi), 93
Liberace, 83
Lifeboat, The, 69
Lifeboat Lodge, 67
Lighthouse Corps, 127
Limelight Department, 113
Limehouse, 69
Lindsay, Vachel, 12-13
Love, Bettie, 11, 31, 53, 120
Love, Richard, 31, 53, 120
Lowry, Terry, 91-92, 135
Lum, William and Mildred, 124

Lum, William and Joy, 97
"Lying in the Gutter," 57-58
Lynch, Thomas, 66
Lytton Springs, 20, 70, 81

McClatchy, Jean, 93
McFee, Joe, 19, 69, 105-106
McDonald, Jeanette, 37
McDougald, Isa, 87
McHarg, Inga, 136
McHarg, William, 83, 125, 136
McIntyre, Charles, 76
McIntyre. Elnora, 88
McKinley, Edward, 43, 56
McMillan House, 68
McNab, John, 44
McPherson, Aimee Semple, 110
Madsen, Harold, 26
Maimonides, Moses, 86
Major Barbara, 111
March of Dimes, 86
March of Witness, 102
Marshall, Thomas, 10
Martin, Dean, 63
Marx, Ernie, 106, 107, 128, 135
Masters, Edgar Lee, 12
Mayor's Office of Emergency Services, 53
Meals That Heal, 118
Mechanics' Pavilion, 36
Men's Social Services, 15, 69
Men's Training Garrison, 18
Methodist Connexion, 5
Mexico City earthquake, 48
Milsaps, John, 15, 18, 19, 20, 39
Missing Persons Department, 23, 82-83
Mission Corps/Community Center, 84, 126
Mississippi River floods, 55
Montgomery, George and Carrie Judd, 20
Morris, Art, 83, 85
Morris, Edith Smeeton, 77, 83, 88
Morris, Henry, 88, 134, 135
Morris, Joan, 88, 135
Morris, William, 83
Morrison, Jane, 95
Mother Teresa, 2
Mount Davidson, 104
Movies with Salvation Army themes and images, 2, 37, 109-111, 113
Mui, Thomas and Joy, 125
Mulpeters, Evelyn, 88, 134, 135
Museum of the City of San Francisco, 35

New Metropole, 70
Newbould, Ardys, 30, 88, 136
Newbould, Victor, 27, 43, 45, 74, 78, 11, 136
New Christian Mission, 3
Newton, George, 14
Nightingale, Agnes, 84
NIMBY, 67, 128
Noland, Joe, 80

O'Connor, Sandra Day, 112
Oakland Hills Fire, 47, 55
Oceanview Corps/Community Center, 126
Old Curiosity Shop, 69
Old First Church, 13
Old Linen Campaign, 39
OMI Business League, 88
Orames, Benjamin, 104
Osbourn, Kenneth and Dorothy, 120

Pace, Rob, 132, 134
Pacific Coast Holiness Association, 14
Paff, Charles, 74
Pardee, George, 38
Park, Justin and Juhie, 127
Park Li, Gimmy, 93, 95
Parker, Edward Justus, 44-5
Parkins, William, 126
Patrick, David and Effie, 67, 78
Pearl Harbor, 44, 81
Pelosi, Nancy, 91
Pershing, John, 39
Phossy Jaw, 1
Pill Box, 67
Pinehurst Lodge, 70
Playboy Bunnies, 95
Ponstler, Robert and Carol, 31
Posillico, Joe, 49, 51, 52, 80, 137
Posillico, Shawn, 78, 137
Prepkit®, 53
Pritchard, Michael, 29, 91
Prince, Russell, 113
Prison ministry, 84
Purviance, Helen, 40

Quakers, 6, 9

Rapson, Della, 74
Rather, Dan, 33
Rainwater, Tom, 80
Ramos, Hector and Gerde, 126
REACH, 118-19
Recycling, 70
Red Shield Clubs, 44
Red Shield Youth Center, 76, 120

Reyes, Moses, 84
Richardson, Lyle, 87
Rigney, Jean, 84
Robertson, Sharon, 80
Robinson, Ray, 32, 48, 53, 117, 136
Robinson, Bunty, 3, 31, 136
Robles, Ana Maria, 50
Rockefeller, John D., 109
Roosevelt, Franklin D. 104
Runyon, Damon, 103, 110
Ruiz, Mario and Maria, 126
Ryan White Care Act, 68

St. Francis of Assisi, 2, 4
St. John, Adela Rogers, 82
Salvation Army bands, 96-98
Salvation Army Canteen Kitchen (SACK), 46-7
Salvation Army World Service Organization, (SAWSO) 52
Salvation Nell, 112
San Francisco City Administration Office, 120
San Francisco Citadel, 74
San Francisco 1906 earthquake and fire, 33-39
San Francisco 1989 earthquake, 52-53
San Francisco Forty-Niners, 112
San Francisco General Hospital, 84
San Franciso Junior League, 70
San Francisco Parades, 102
San Francisco Service Center, 100, 108
San Francisco Song Festival, 13
Santa Lucia Festival, 111
Sarah, Duchess of York, 87
Saunders, Rob, 29
Scandinavian Division, 26
Sears, 121
Senior Meals and Activities Program, 118
Senior Picnic, 118
September 11, 53
Service Club Bell Ringing Day and participating clubs, 95-96
Service Extension, 44
Shaw, George Bernard, 2, 98, 111
Sheldon, Margaret, 40, 42
Sherman and Clay, 97
Shennan, Minnie Belle, 77
Shiroma, Larry and Vickie, 67, 127
Sholin, David and Deanna, 127
Shultz, Charlotte, 3, 90-3
"Silver Bells," 109
"Silver Hill," 75-6
Silverberg, Rebecca, 96

Silvercrest Residence for seniors, 117-8
Simmons, Jean, 2, 109
Simpson Bible College, 76
Simpson's Pot, 105
Sister Stuyvesant, 35
"Skeleton Armies," 101
Skid Row, 63, 66, 128
Skid Row wino glossary, 59
Smith, Frank, 18
Smith, Lawrence, 81
Smith, Lawry, 67, 127
Smith, Phil, 67, 68
Smith, Wilma, 88
Soup kitchens, 6, 112
Sousa, John Phillip, 2, 100
South of Market, 14, 116
South of Market addresses where the Army has worked, 123
Sparks, Les, 67
Sri Lanka, 14
Staley, Geneva Ladd, 40
Stanton, Elizabeth Cady, 11
Starr, Kevin, 4
Stead, W. T., 8
Steele, Clinton, 118
Steele, Dorothy, 84
Stidger, William L, 12
Stillwell, Henry and Mary, 15, 17, 18, 19
Storey, Joye, 55, 100
Strickland, Ron, 97
Submerged Tenth, 8, 12
Sundin, Ruth and Wes, 127
Suzuke, Naoji., 81
Sutro, Adolph, 69
Swig, Ben, 3, 90

Taliaferro, Ray, 81
"Take Charge Your Life," 82
Taverna, Angelo, 96

Tenderloin, Housing and Community Center, 128-129
Thanksgiving Half Time Extravaganzas, 112
The Theodora, 19
Theroux, Mary, 92, 132
This Little Light of Mine, 87
Thompson, Francis, 9
Thrift Store Trucks, 69
"Thursday Night at the Army," 111
Tikkun Olum, 86
Timpson, Neil and Kathy, 129
Tobin, Doris, 80
Tollerud, Sheryl, 111
Tom, Sue, 124
Tournament of Roses Parade, 100
Toy, Ron, 28, 47, 86, 126
Toy, Keilah, 28, 126
Tracy, Spencer, 37
Training College for Officers, 75, 76, 78
Training College Parade of Life-Saving Guards
    and Guard's Divisional Drum Corps, 79
Training College and Northern Division
    Annual Band Festival, 79
Tzedakah, 2, 86
Turk Street Corps/Community Center, 127-129
Twain, Mark, 2, 34-35

USO, 44-45
Udy, Nancy, 100
Union Square, 14, 36
Union Square Association, 14, 92, 94
United Anglers Enforcement Committee, 87

United Crusade/United Way, 83

Valencia, Javier, 93, 99
Volunteers of America, 116

Wahl, Jan, 93
Wave, The, 22
Washington High School Key Club, 95
Waste, William, 132, 134
Watson, Robert A. 82, 131
Wells, Alfred, 14, 15, 17, 18
Wells, Polly Medforth, 17, 18
West, Mae, 2
Weston, James, 84
"While Women Weep," 9
Willard, Frances Elizabeth, 10
Williams, Laurie, 30
Wisbey, Herbert, 43
Wood, Sam, 35
Wolf, Jack, 120
Wolf, Virginia, 118
Women in The Salvation Army, 9-11
Women's Auxiliary, 30, 88, 135
Women's Lib, 10
Women's Training Garrison,
Working Man's Institute,
Wun, Daniel and Susan, 125

YMCA, 43, 87
YWCA, 44, 87
Yamamuro, Gunpei, 80
Yee, Check, and Phyllis, 28, 124-126
Yepez, Maria and Pedro, 126
Yerba Buena Corps, 118
Young, Stella, 40
Youngquist, Lucille, 88

Zodiac Killer, 80
Z95 Radio, 94